## *Other Books by the Author*

*Introduction to Environmental Physiology.* 1966.
(Reprinted 1969)

*Textbook of Environmental Physiology, $2^{nd}$ Edition.* 1974.
(Reprinted 1985)

*Principles of Integrative Environmental Physiology.* 1998.
Co-authored with Marvin L. Riedesel and Diana L. Thrift.
(Reprinted 2000)

# SCIENCE ON THE FAR HORIZON

G. Edgar Folk

*with*

Diana L. Thrift

INFINITY PUBLISHING

Cover photo of man on dog sled, 1961, Barrow, Alaska, by author
Cover photo of author by Michael D. Henry

Unless otherwise credited, all photos were taken by the author, G.E. Folk, or are from his personal collection.

Parts of Chapter 6 appeared in the article, "The Harvard Fatigue Laboratory: contributions to World War II," published in Advances in Physiology Education, 34:119-127. Folk GE w/ Thrift DL, 2010.

ISBN 978-0-7414-7899-3 Paperback
ISBN 978-0-7414-7875-7 Hardcover
ISBN 978-0-7414-7876-4 eBook
Library of Congress Control Number: 2012917899

Printed in the United States of America

Published November 2012

INFINITY PUBLISHING
1094 New DeHaven Street, Suite 100
West Conshohocken, PA 19428-2713
Toll-free (877) BUY BOOK
Local Phone (610) 941-9999
Fax (610) 941-9959
Info@buybooksontheweb.com
www.buybooksontheweb.com

*For my daughter, Victoria Elisabeth.*
*You provided a constant source of inspiration.*

# *Contents*

# *Foreword*

THE WORLD—THE world of primal forests in blossom and bloom, the world of a bald eagle soaring, a bear emerging from hibernation, and butterflies of so many different kinds and colors—as a child, I encountered all that, seeing it all through my father's eyes. To his students and colleagues in the scientific community, my father was looked up to first and foremost as a teacher and research biologist. But to me, he was this wonderful man with binoculars and a baseball hat, who made time to introduce me to the wonders of nature, making each adventure with him a new horizon to behold.

### *Spring*

I am seven years old and running after my father as he climbs to the top of a small mountain in the woods of New Hampshire. I am breathless when I get to the top, and my father is smiling and pointing to the horizon.

"Look, Butch, do you see what I see?" he says, pointing excitedly. "That is an American Bald Eagle flying above us. Isn't it glorious?" I look up to where he is pointing, and I don't see the eagle. But I look at my father's face, his infectious enthusiasm, and don't dare disappoint him. "I see it! It's beautiful," I say.

**G. Edgar Folk Jr, Victoria, G. Edgar Folk Sr. Andover, MA, c. 1947.**

I remember that same spring when we find, in the forest, a lady slipper—a delicate orchid-like flower that is indigenous to Maine. He points out the white and pink blossom and tells me it is typically found in the undergrowth of dry, mixed woods and boggy soil. I am in awe of the beauty of it as seen through my father's eyes, and the memory of the lady slipper remains with me to this day.

### *Summer*

I am four years old and "brown as a berry," as my father loves to recall. We are on Naushon Island off the coast of Woods Hole, Massachusetts. That summer with my father, we explored every nook and cranny of this strange island. The terrain was rocky and filled with saltwater pools, and I learned about starfish and mussels, barnacles and kelp, insects and (to my innocent eyes) exotic sea birds. Each day was full of surprises as my father stopped often to teach me about the flora and fauna of the land. When I was tired, he would hoist me into his backpack, where I stood precariously; I would put my small hands around his forehead, and we would continue our travels over the island. Once again, I see all the wonders of nature through my father's eyes.

### *Autumn*

I am nine years old. We live in Maine, and my father is a professor of Biology at Bowdoin College. On weekends we explore the woods, going to a different place each time. On this occasion, I am told by my father that I am hiking in a "virgin pine forest," and I breathe deeply to remember the smell of the decaying pine needles covering the forest floor. My father is pointing up at the pine trees.

“These are very old trees,” he lectures, and again I listen attentively.

I question, “How do you know?”

He says, "You can tell by the height of the trees, the lack of branches at the bottom, the top branches reaching up to seek the light."

I am amazed at how much he knows. Then, for a moment, I find myself alone as my father goes off to collect "a specimen" for his biology students. I busy myself making a path of pine needles on the floor of the forest. I hear him call to me and I run towards his voice.

"Look, my darling," he says, holding out the specimen. It is a large blue jay and the bird is dead. He guides my hand to let me feel the feathers. I am saddened and he tells me not to worry. "The bird is sleeping peacefully now," he says softly. "Let's find him a resting place." He helps me dig a little grave, and we place the bird gently in the ground. I cover the grave with pine needles and rocks. Would it be there the next time I returned?

### *Winter*

How often do we begin to understand our childhood experiences in light of the reality not only of our own life but also of a father's life?

I am ten years old and trudging through deep snow near a ski resort in New Hampshire. My half-brother, Kit, is tall and can keep up with my father. I lag behind. We reach the pinnacle of a cliff and we stop, listening to the sounds of the wind through the snow. My father is scanning the valley below with his ever-present binoculars.

"Kit, Victoria," he says. "Look at that tiny black speck below."

My brother takes the glasses from my father and deftly puts them up to his eyes. "Yes," he says. "I see it. Is it a black or brown bear?"

I excitedly grab the glasses away from him, anxious to see a bear, and I use one side of the binoculars as a telescope, scanning the valley below for the moving black speck.

"Dad!" I scream exuberantly, "I see it! I see it!" I had only caught a glimpse, but I had seen it. I look up into my father's eyes as he beams with approval. Later, I brag to my friends that I have seen a big black bear while hiking with my father in the snow.

### *The Seasons to Come...*

Looking back, I know I yearned for more times like these, more time with my father, but like many, he worked too much, and I missed him terribly. Yet he remains to this day my hero and my teacher. Remembering the rites of passage, learning from my father, has been an experience bar none—an almost religious experience, an ongoing outlook on life for years to come.

In 2012, my father is entering his 98th year. The laughter, the memories, the profound respect for his dedication and contributions to the world of science and his stories, his delightful stories, are but a part of how I see this charming man, G. Edgar Folk. How blessed I was to have this man sharing his world with me.

Now, you have the privilege of sharing his world, too!

*Victoria Folk Sprague, Esq.*
*May 2012*

# *Introduction*

THE INSPIRATION FOR recording this personal history came from a desire to share the excitement of being a field scientist not only with those whose lives have followed a totally different pathway, but also with my fellow travelers as if we were sitting around a fire swapping stories. Where does the energy to write a memoir come from? From a desire to describe and share one's experiences before these memories fly away like a flock of birds.

In writing this memoir I had several ambitions. One was that I wanted this book about "a life in science" to be accessible to a broad audience of people interested in nature, science, history and biography. It was not written to be a scientific treatise, although one will find general accounts of research studies as well as descriptions of reptiles, mammals and the environment. (Readers desiring more detail on a particular subject can use as a jumping-off point the references at the end of the book.)

Because I was privileged to learn from and work with some of the "greats" in science, I have endeavored to describe and pay tribute to these exceptional individuals. I also wanted to set down a record of the key people, places, and events that led to the development of *Environmental Physiology*, a new discipline within the field of Biology.

I have been fortunate to have lived through most of the last century; thus, I have tried to give a sense of the times in which I grew up and the many changes I have witnessed. These include horses and buggies making way for motorcars; communications evolving from telegraph to telephone to email; lab books giving way to computers; and in the Far North, dog sleds being overtaken by snowmobiles.

In the first half of the twentieth century, the world seemed bigger and more mysterious; new frontiers beckoned from all directions. There was a feeling that everything was possible if you just worked hard enough...and had some luck. World War II affected us all in so many ways, disrupting plans and requiring sacrifice, redrawing borders and alliances, and advancing medicine, science and technology to propel us into new dimensions.

This is a memoir and as such is subject to memory. Many events are matters of historical record, others are my recollections; readers can be assured however, that we've done our best to ensure accuracy throughout.

I hope the story of my life in science will inspire readers to pursue a career that captures their hearts, be it on horizons far or near. In writing about the many hurdles I had to jump to follow my passion, I want to underscore the value of perseverance. In my 98$^{th}$ year, I still believe that with enough hard work—and a bit of luck—anything is possible. In the words of Thomas Jefferson: *I'm a great believer in luck, and I find the harder I work the more I have of it.*

And finally, if any of my readers are motivated to pursue a career in science, I would be well pleased.

*G. Edgar Folk*
*Iowa City, IA, April 2012*

# *Prologue*

## *Class of 1940*
## *Harvard Graduate School*

### Combining the Outdoor Life with Research

IT WAS 1939, THE SAME year Germany invaded Poland, and two years before the US would enter World War II. In the basement of the Biological Laboratories of Harvard University in Cambridge, Massachusetts, a group of graduate students was playing volleyball, as they did nearly every day.

There were no specific teams in these informal matches; each person entering the game joined whichever side had fewer players. I was one of those students, and the reason I remember these games is that most of the players went on to become eminent biologists, despite the four-year interruption of WWII, which caused most of us to postpone completing our graduate work until the war was over.

The players included Donald Fawcett, Donald Griffin, Oliver Pearson, George Bartholomew, Peter Morrison, Carol Williams, Nevin Scrimshaw, William Carroll and Ralph Smith—names that will be recognized by many of today's scientists and those interested in the biosciences. After a rousing game, we were apt to eat lunch together to discuss our common interest in field work, and ways to combine an outdoor life with research. We often gathered around a toaster and peanut butter jar in the large laboratory I shared with Ralph Smith four floors up from the volleyball room. Collectively, these students formed an intellectual stew pot with much exchange of ideas and even heated arguments. I was fortunate to be a part of this particular group; we shared with each other the excitement we experienced in our laboratories and in the field. We inspired each other.

To gather experimental material from where the evergreens were heavy with snow—this was what we liked to do. We all wanted to do as much of our research as possible "on the far horizon." For this reason, Ralph Smith captured crayfish from the Charles River for his thesis work when he could easily have purchased them. It explains why George Bartholomew had to sit out some volleyball games: he'd fallen from a ladder one night while catching English sparrows for his studies. In his pursuits, Peter Morrison brought wild shrews into the laboratory to measure their metabolism, and I collected hibernating bats from a cave in NY and brought them to our cold room so I could study the relationship between their reproduction and hibernation. Donald Griffin crawled through hundreds of dank caves and abandoned mines attaching tiny numbered bands to the legs of hibernating bats so he could study their migration. Don had inspired many of us to follow him into these caves to work alongside him—and above or below him—to assist in his quest to band every hibernating bat in New England.

Don Griffin was typical of those scientists whose desire was to carry instruments into extreme environments and study life there, in addition to bringing back acclimated animals to study in the laboratory. Later on, all of us would carry instruments to locations as far flung as the Caribbean, Europe, Canada, South America and the Arctic to pursue independent studies of insects, birds and mammals.

This field would later become known as *Environmental Physiology*. I have worked in this area for my entire career, and played a role in naming and defining the discipline.

Scientific research, whether conducted in the field or in the lab, can be exciting—and at times even dangerous. Oliver "Paney" Pearson learned this while still a graduate student at Harvard, where he was working late one evening. When I stopped by his laboratory, Paney greeted me enthusiastically.

"I'm certainly glad you dropped in Ed, because I have to talk to somebody. I just did an experiment and the results are hard to believe. This weekend I trapped some *Blarina* shrews—you and I had talked about how curious it is that such a small animal could kill animals larger than itself for food. So I thought I would test their saliva. I ground up the salivary glands with sand and made an extract."

He pointed out that naturalists were always speculating about the possibility of poisonous saliva. "*And really Ed, to disprove them, I was tempted to inject it under my skin. Instead, I injected it into white mice.*

*They all died instantly. Now I've worked all evening trying to find what to compare this extract with, and I found that the venom from this shrew was drop for drop as poisonous as that of a Cobra."*

Paney quickly published his results in a beautifully written and detailed paper (Pearson 1942). His experiment with the shrews prompted me later to test the saliva of so-called non-venomous snakes in a similar way, with the same results.

All of these budding *Environmental Physiologists* had something else in common, an inspiring professor as a mentor. Many who came to graduate school were encouraged by the great Professor Laurence Irving of nearby Swarthmore College. Larry Irving kept a manatee in the college's swimming pool where he and his team could study its diving mechanism and thus, Don Fawcett was able to pursue his own investigations of the unique anatomy and physiology of the Florida manatee. Others of us were inspired and mentored by John H. Welsh, discoverer of *serotonin*, and later by Bruce Dill—the consummate "scientist's scientist." I will have much more to say about these two eminent scientists in later chapters. Nobel Laureate George Wald was our Biochemistry Lecturer and he was the specific reason Bill Carroll had come to Harvard; they soon published together on Vitamin A and carotenoids. Prof. Frederick Hisaw discovered the hormone *relaxin* in his laboratory, and motivated Oliver Pearson to later scale the Andes looking for this hormone in wild ancestors of the domestic guinea pig.

Two of the staff members who joined our volleyball game were Drs. Horace Drury and Herman Rahn. Horace Drury taught us embryology at Harvard, and later moved to the Arctic where he studied survival under extreme conditions and directed the Arctic Aeromedical Laboratory in Fairbanks. Herman Rahn was a post-doctoral fellow in the Harvard Biology Department, and would become a frequent passenger on Scholander's research ship, the *Alpha Helix*. Rahn was such an influence on me that he later persuaded me to collect material for him in remote mountains on the island of Haiti.

In addition to their countless investigations in the field, this volleyball group went on to successful academic careers. Carol Williams became a biology professor at Harvard; Nevin Scrimshaw became a professor at MIT; Donald Griffin was elected to the National Academy of Science; ballplayers Ralph Smith, Oliver Pearson, and George Bartholomew all became department heads at large universities; Donald Fawcett led the Department of Anatomy at Harvard.

The graduate students in that group had many things in common, not the least of which was our passion for the natural world. I'm sure the others must have been asked often, as I was: "How did you get interested in biology?" I suspect that, as in my own life, their interest in biology began early and was fostered by family, teachers and colleagues.

Therefore, using myself as one example, I will push back the clock to describe some of my earliest experiences in the making of a biologist.

# *Chapter One*

## *The Making of a Biologist*
## 1914–1932

### Boyhood

TO MY YOUNG eyes, the far horizons began at the brook behind our garden in the small town of Andover, Massachusetts, where I lived from the age of six. It was a large garden with fruit trees extending to the brook, which was inhabited by trout and later by two wild Mallard ducks that I raised. These wild ducks lived in the brook, but when I came out on our back porch and rattled corn in a pan they flew out of the brook to land at my feet. Beyond the brook, there were extensive pastured hills upon which I learned to ski.

The locale where our house was situated seemed designed to inspire a person to natural history. We lived on Chestnut Street, where the large trees lined the street and their branches formed a canopy overhead. I could reach dense woods from our house by two pathways: one was to walk up the hill on Chestnut Street, but the best way was to walk along the brook that formed the back boundary of our property.

I was the youngest of three children. I had two older sisters, Evelyn Alice and Barbara Sylvester (Barbara got her middle name from my mother's favorite doctor). As the only son, I replaced my parents' firstborn, a boy named Grosvenor who had died at the age of three from scarlet fever. My mother, accordingly, was overprotective, but my father made up for this by encouraging my outdoor adventures and constantly reminding me to work hard.

My father was a Methodist minister who embodied the role of curate-naturalist as described in the literature of Charles Dickens. He grew up on the shore of Chesapeake Bay as a typical "barefoot boy," one of eight children in a family of all boys. His father maintained track for the railroad, and this did not pay sufficiently to provide food for such a large family. They lived in a wooded area and to a certain extent, lived off the land. By the time I was six years old, my father was teaching me how to make a wooden drop trap to catch rabbits, which in his early days had provided food. My father had been an avid hunter known for his ability to shoot a squirrel in the air as it jumped from one branch to another. He still had as a curio an antique flint-lock Chesapeake Bay goose gun that had a long history in the family. I eventually became the keeper of that very tall goose gun, and a few times when it was being moved, the tip of its barrel punched a hole in my ceiling. I later gave it to the Audubon Society for its museum.

**G.E. Folk Sr., WWI.**

I was born in 1914, right at the beginning of the First World War. Although my father was older than most recruits and was not required to serve, he felt that, as a minister, he should be there for the "boys" who had to fight. He enlisted, and was gone from our family until after the war's end in 1918. World War I was called the war to end war, sadly proving to be over-optimistic.

I was named after my father, George Edgar Folk. Thus, I inherited a first name I would never use—or even know how it got in the family. My father felt sure there were Georges somewhere in the line, but he wasn't sure where or when. He went by the nickname "Ted." I had the misfortune to have a mother who called me "Junior." Everybody hated it—including me—but she persisted, and no manner of objections on my part could change her mind. Not, that is, until I was in 9th grade. One of my teachers at Andover asked everyone to tell him their nicknames.

"I don't have one," I replied.

"Everyone has to have one," he decreed, and decided for me right there. "I'll call you Ed."

I was elated, and told my parents that evening. Mother didn't like it of course, and was even offended, but Dad was relieved. He threw his weight behind "Ed," and from then on, that's been my name.

My mother, May Davis Folk, was born "Minnie May," but she disliked that name, and always went by "May" (the irony in this she apparently never noticed). She grew up the daughter of the man who sold and delivered milk to most of Boston. They had a large stable beside their big house, and my mother learned to drive a horse and buggy at an early age. My grandfather also built a large brick factory to make ice cream and opened several ice cream parlors. He required that both his daughters wait tables in them while they were growing up.

**Hitched and ready to go. Grandfather's milk delivery wagons.**

I never knew my grandfather, but I know he was proud of his horses and his business. He was also apparently very competitive. Family legend has it that if my grandfather was driving his horse and buggy down the street with his horse at the trot, and he was passed by another horse and buggy, then my grandfather would have to go out and get himself another horse who could trot faster.

Our mother said that she always wanted to marry a minister, and she did. She was a strong-willed woman who served on the school board as I was growing up. She was active in book clubs and other activities, much to the chagrin of my sisters, who had to take over the house cleaning.

Automobiles had begun sharing the streets with horses and buggies, but we didn't own a car until I was eighteen, so we walked nearly everywhere we went. I clearly remember bright red fire engine wagons drawn by white horses that galloped through the streets—I think most fire-engine horses were white so they could be easily seen.

Each day of the workweek, my father walked to the center of town to take the streetcar to the city of Lawrence, Massachusetts where he was a labor arbiter for the American Woolen Mills. He settled differences between labor and management when there were strikes, which several times had brought the city to a standstill. Indeed, thanks to these successful arbitrations, at least twice when I was a small boy, I was stopped on the street in Lawrence and embraced by people who exclaimed, "Your dad is a great man!"

My father also worked for the YMCA organizing classes in English for the Eastern European immigrants who worked at the mills. He often brought these students home for dinner. One, a refugee named Jules Detweiler, claimed to have been at the Tsar's court in Russia where he knew Rasputin, the "Mad Monk" who influenced the last of the Russian royals. Jules told us he had escaped from Russia by hiding in a load of hay. He became a good friend of our family, and even took my sister, Barbara, on several dates.

My father was well known in the area not only because of his job as arbitrator, but also because on Sunday mornings he could be found at the pulpit in this church or that one, as he had no fixed congregation. He was a traveling minister, filling in for vacationing or ill pastors in protestant churches around Andover and Lawrence.

He preached in the "fire and brimstone" style. He also spoke out against local gangsters. This disrupted the family when some of the gangsters threatened to do us harm if he kept it up. His response to that was to go out and buy a six-shot revolver and take my poor mother out in the yard to do target practice. She had led a protected life, and was the last person in the world I would have expected to be shooting a gun. At the time, I was too young to realize what was happening, but later my older sister Barbara filled me in, showing me the six-shooter hidden in my dad's top bureau drawer.

## *Nature Walks*

Most Sundays we walked in the woods, taking our dog, Buck. He was a Boston Bull Dog whose short legs still managed to run circles around the rest of us.

We walked in all four seasons when the weather was agreeable. My mother always accompanied us, despite her bad back, which kept her from participating in other sports like tennis. Both of my parents enthusiastically pointed out the spring flowers and the evergreen plants that always survived the winter. I could provide quite a list of the plants that became important to my two sisters and me, but the outstanding ones in the spring were Jack in the Pulpit, Dogtooth Violet and Shooting Star.

**Ed and Buck.**

My father's love of plants carried him to a hobby of hybridizing Gladioli, and eventually he became President of the International Gladiolus Society. My mother specialized in the wild flowers; we dried them out between pieces of paper and mounted them museum style.

However, it was the animal life my father pointed out that eventually led me to my career. He had some knowledge of birds, and we would steal up softly to watch a drumming ruffed grouse. My father also liked to point out the larger songbirds. Most important, he would turn over logs and rocks to see what lived underneath. I will never forget seeing my first spotted salamander, a startling animal that can grow to be five or six inches long. In other places he pointed out *Plethodontid* salamanders—thin, elongated creatures that do not have lungs.

Of my two older sisters, one of them was an outdoorswoman and athlete, while the other had a sharp mind. Barbara, the athlete, taught me early to be fearless when sledding down a steep hillside, or bobsledding on an icy street. Winters were often quite severe, and snowshoes were often a necessity. When snowdrifts covered the fences, Barbara and I skied right over them in our toe-strap skis. Because of this early training, I would later ski down mountains.

**Folk Family Portrait.** ***l to r:*** **Barbara Sylvester, G.E. Jr. (Ed), May Davis, G.E. Sr., Evelyn Alice.**

Evelyn was "the brain." I don't think she ever got any grade other than an A in school, and it came as no surprise when she graduated *Phi Beta Kappa* from Boston University. It was my father's alma mater for his divinity degree, and he was understandably proud of Evelyn's record.

## *The Abbott Homestead and the Shawsheen River*

Early in our life in the small town of Andover, we established a friendship with the Chester Abbott family. These Abbotts were one branch of the large and well-known family who were among the first European settlers in Andover and who had founded the exclusive Abbott Academy for girls. Our Abbotts ran a dairy farm, and on it was one of the oldest houses in New England, the Abbott Homestead, built in 1685. The family lived in another farmhouse nearby, but maintained the homestead for the Historical Society.

Mrs. Abbott was in charge of this famous house, and by appointment showed it to people interested in early history, which included "Indian Walls," witchcraft accusations, and abolitionist meetings. The Indian Walls consisted of sliding wooden panels that could be pulled across the windows during Indian attacks, and fortified walls that had a layer of bricks between inner and outer sheathing, all covered with a mixture of plaster, clay and straw.

It was about a 30-minute walk from our home to the Abbotts', and our family was frequently at the Abbotts' farmhouse. I went inside the old homestead only a few times, preferring the outdoor explorations the place offered. On the lawn stood the locally famous 250-year-old elm, and across the estate ran a long, winding esker (a glacial landform) known as Indian Ridge.

The Abbotts' son was called "Chick," and he was my best friend. Chick had two horses and taught me how to ride with an English saddle and bridle. We rode through the miles of forest adjoining their farm, down country lanes, across pastures. Galloping along Indian Ridge, my mount unseated me; unhurt, I climbed back on and we continued our ride. These outings spurred a passion for horses that has never left me, and for the rest of my life I tried to ride as often as my state of affairs would allow.

**With one of my feathered friends.**

Mr. Abbott operated the farm, which included the large fields in which I earned my first dollar. At the age of eleven, I found myself on hands and knees in the Abbotts' cornfield. This was due to my father's constant reminders as I got older of the necessity to earn money to help pay tuition. Consequently, my first job was thinning corn for Mr.

Abbott at fifteen cents an hour, six hours a day. Thinning corn involved crawling down the cornrows where each cluster had about six sprouts. I was to pull out the three weakest ones and leave the three strongest, most viable sprouts. Fortunately for me, Chick crawled along beside me and we talked excitedly during the long, hot hours. At the end of the six hours, I was then to herd calves grazing on the lawn of the historic homestead. When the time came to be paid, I calculated ninety cents. Mr. Abbott handed me a dollar bill; I protested that it was supposed to be ninety cents. He reminded me I had forgotten that I tended calves, which would add up to one dollar. This was the first dollar I had ever earned.

Beside the Abbotts' cornfield flowed the beautiful and extraordinarily wild Shawsheen River. This river played an important role in my childhood, not only because of the adventures it offered, but also because of its influence in shaping me into a field biologist.

The Shawsheen River was small but deep with a mostly sandy bottom and a width of thirty feet, although at times it flowed through portions of marsh. *Shawsheen* is an Indian word said to mean "Great Spring," and the river in fact originated from a spring rising from a field in Bedford, about twenty miles south of Andover. This remarkably clear river flowed northwards, winding its way through woods and agricultural fields before arriving in Andover.

Farther downstream it joined the larger, historic Merrimack River. The Merrimack was famous because for several years after the landing of the Mayflower, many of the ships that sailed from England for the New World discovered that the mouth of the Merrimack made a good harbor. One of the diaries of that time reported that the tall masts of so many ships anchored there one year made the mouth of the Merrimack look like "a forest."

### *A Very Useful Canoe*

With its proximity to both the Shawsheen and Merrimack Rivers, Andover had once enjoyed a very active canoe club; across the river from the Abbott Homestead stood the club's canoe house with a ramp to the river. It was wonderfully situated—to the North was the confluence of two rivers, and to the East on a neighboring farm there was a deep lake with an entirely sandy bottom surrounded by sandy banks.

This deep body of water was called Foster's Pond, and although privately owned, it was a favorite swimming hole. The canoe club, with astonishing ingenuity and energy, had dug a narrow canal from the Shawsheen River to this pond.

When my father learned about the river that ran through the Abbotts' property, he purchased a fourteen-foot *Old Town* canoe to ensure that he and his family could explore these pristine waterways. This canoe was the nearest we came to owning an automobile for many years.

From then on, there were fewer walks in the forest as we turned towards the river. We would start upriver as early in the year as Washington's Birthday in February; my father in the stern, older sister in the bow, and my mother, other sister, me and Buck sitting in the bottom of the canoe. That dog was a good sailor! We would paddle upstream for half a day, eat lunch, and paddle back to the boathouse. We did this from the time I was about seven.

As we navigated the Shawsheen, we came to know some of the river's abundant aquatic life. We saw four species of turtles, including numerous large snapping turtles, huge bullfrogs and other species of frogs, and extremely large fish, especially pickerel and yellow perch.

From about the age of twelve, I began to paddle out in the canoe by myself, though of course I let Buck come along. That little guy had fun in the canoe, especially the time we scared up a great blue heron—Buck felt it was his duty to jump out of the canoe and try to catch it, despite that the bird was bigger than he was and airborne. He didn't catch the heron, but made up for it at home, leaping into the air to catch the rope my father had hung from a tree so my friends and I could climb. He would catch the thick, knotted rope and then hang on, fighting and growling ferociously while he dangled above the ground.

Steering the canoe along the shore and using my collecting net underwater, I could achieve a bounty of fascinating critters. There were baby eels, which eventually would grow up to make their way by river to reach the sea as adults. There was an especially interesting small fish called the Johnny darter, which didn't have a swim bladder and stayed at the bottom of the river hopping from one location to another. All the fish were egg layers, and there were numerous babies. The long, shallow canal from the Shawsheen to Foster's Pond was the ideal place for baby turtles to spend the initial few years of their life. I estimated the canal to be about 300 feet long because it took me about 100 strokes to reach the pond from the river.

The gallon jar I carried on my bicycle transported attractive baby blue sunfish, baby pickerel, and yellow perch, which I used to fill four aquaria in our home kitchen. I traveled by streetcar to pet shops in Lawrence to buy angel fish and other exotics and soon found that the owners were fascinated by all of my catch in the river. Before long, I was carrying my produces of the river in one direction and coming back with what I considered beautiful tropical fish in exchange. This went on for years, and was especially valuable later for one of my jobs at Phillips Andover.

## Preparatory School: Lifetime Influences

I'd been born in Southborough, MA, but not long afterwards, my father had moved the family north 45 miles to Andover because of its schools. The girls' Abbott Academy, and the boys' Phillips Andover Academy, were both known for their high scholastic standards. My father's salary from his several jobs was not adequate for paying tuition for the girls to attend Abbott and for me to attend Andover, thus, from early on, he instilled in us the need to help with the costs.

"Son," he often told me, "if you want to go to the prep school, you will have to work to earn money."

### *The Financial Effect*

This leads to a theme I'll call *The Financial Effect.* As I look back, I realize that the need for me to earn money to finance my education lasted all the way through graduate school. Growing up, my sisters and I were supposed to work summers to pay part of our tuition, and that had been the push for earning my first dollar in the Abbotts' cornfield. My sister Evelyn and I both had many jobs while growing up, but Barbara somehow managed to spend her summers going to camp with her friends—this engendered many family disputes, especially between "the athlete and the brain."

At age thirteen, with my sensible father's prodding, I obtained a summer position working for a caterer in the industrial town of Lawrence. We traveled by truck 30 miles to Fort Devens to wait tables for the officers of National Guard units who had to train there during the summers. Each year for five years, as soon as classes ended for the summer at Phillips Academy, I went to join the catering crew and worked until September 1st. We worked as many as ten hours a day, seven days a week, serving meals to large groups of officers. The work was difficult, and my uncle Claude Davis, a successful, hard-working businessman who had disapproved of the way my mother tried to spoil me, later announced: "They made a man out of you."

Working at the military camp gave me the opportunity to visit the large, mile-square parade ground. I loved to stand beside the General's reviewing stand to watch the complete Fort Devens Reviews. The first two summers, with the army still horse-dominated, columns of cavalry and horse-drawn caissons passed in front of the reviewing stand. The columns began to merge as they passed until finally, there was a mile-long line of tossing horses' heads extending at a right angle from the reviewing stand. Meanwhile, overhead, single-engine one-seater military biplanes passed close to the reviewing stand. Over the five years I worked there, I saw the army that paraded in front of me change from columns of cavalry and horse-drawn caissons to columns of jeeps, tanks and motor-driven caissons.

During the school term I had student jobs along with my studies. As a work-study student I could also help defray the cost of tuition if I maintained adequate grades. Thus, I learned early a principle that was to follow me through my many years of studies: grades mean money. Andover had a very simple formula for a work-study student. If you had a B average, you paid no tuition. If you had an A average, both your tuition and room and board were paid. Unfortunately for me, Evelyn had grabbed more than her share of brains in the family; when I could get my average up to a B, my family was more than grateful.

My first position had to do with books and I loved that job because I was already addicted to them. On the campus at Andover there was a Gothic library building (some people called it "monstrous"). In my first year they built what was called the Oliver Wendell Holmes Library, which was a magnificent Georgian structure. I was a "stevedore," joining the crew whose job it was to move the books. With two-handled wooden trays, we plodded from the old library to the new, deposited the books from the tray, and went back to the old library for more.

Meanwhile I was struggling with Algebra, English, Latin, French and History. This program went on for years.

### *A Sampling of School Culture*

My mother's cousin, Guy Eaton, was an instructor of mathematics at Andover, but more than that, he was an interesting influence on my entire life. He took me over as a shy, pudgy first-year student and taught me not only Algebra, but also social skills. He was in charge of extra-curricular activities for the school—and he urged me to choose one and participate.

Which school club to join? I had learned to play jazz drums, and joined a group at Andover called The Riveters. This group didn't last very long because the demands of coursework became too great.

The leader of the symphony orchestra, Carl Phatticer, urged me to be his timpani player; thus I played timpani for four years. I was very fond of the fascinating Mr. Phatticer, partly because of the carillons, a beautiful set of chromatically tuned bells that hung in a tower on campus. I was honored to climb the tower stairs to the keyboard and stand beside the professor as he played the carillons. Perhaps it should not be called a keyboard because it actually consisted of levers, and Phatticer would push with his fist, lever after lever, producing beautiful songs that could be heard over the entire campus.

As head of extra-curricular activities, Guy Eaton had arranged for the Glee Club and Orchestra to travel and give concerts at the female academies and even Colby Junior College. The concerts were always followed by dancing. Fortunately, my mother had previously forced me to take dance lessons. But I shouldn't complain, because now, in my nineties, I still like to go ballroom dancing!

Guy Eaton was also the person who insisted that I become an usher in the school's auditorium. This required wearing a tuxedo, and of course my mother was delighted. Somehow my parents managed to find the funds for the tuxedo, and after class I had to rush home to change and hurry back to school dressed in the tux. The ushering was not only satisfactory, but also important to my interests. There were fine musical programs, one featuring the great Opera star Schuman Heike. It was on such nights that I had to wear a tuxedo. Other nights for the presence of famous explorers, no tuxedo was needed. I especially remember showing slides for Roy Chapman Andrews, explorer of the Gobi Desert, and for the most exciting person of all, Commander Byrd, who talked to us about Antarctica. It was my privilege to serve a glass of water to these great field workers before their lectures.

It was also in Mr. Eaton's Algebra course that I became friends with a student from Hawaii named Sargeant Kahanamoku. He couldn't seem to grasp the ideas of Algebra so we met for 30 minutes before class each day and I went over the exercises with him. Meanwhile, he was breaking every swimming record at our school. To this day, I suspect that Kahanamoku's swimming accomplishments are still on the Board at the pool. Later, the United States printed a postage stamp in his honor. Kahanamoku only stayed at Andover a year and then went back to his royal family in Hawaii, where he went on to become a surfing legend.

### *Essay Questions and "Good Writing"*

An English teacher, Graham Benedict, always gave tests with essay questions, which I credit with both improving my ability to express myself and making me a better writer—but not in the way you'd expect. I'd enjoyed responding to Benedict's essay questions and initially that was reflected in my test grades, all A's and B's. However, when the final course grade was given, I got a C! I went directly to Mr. Benedict, expecting he would correct this error, when he told me firmly, "Your grade will remain a C until you improve your handwriting!" This news was not only unwelcome but also surprising, as I had never been told until then that my handwriting was bad.

I had been taught by the Palmer method, which was how most children learned cursive writing at that time. In this method, one drew circle after circle on a page, and then more circles—possibly to get the flow going—and focused on using the muscles of the arm and shoulder rather than those of the hand.

As soon as I looked at my papers, the problem became obvious: my writing had loops all over the place going in different directions—the loops weren't even consistent within each word! I went home that night determined to change my handwriting. I modified the way I wrote, taking out all the loops, and replacing them with straight lines. In my next essay for Mr. Benedict, there were no loops, and my grades went up. Even today, there are no loops in my handwriting. I shouldn't really fault Palmer's method, but for some reason, I had not learned the correct way to write until that point.

## *Biology, Birds and a Conservatory*

I took a general science course my first year, and had a most inspiring teacher named A.B. Trowbridge. It was his attempt to interest us in outdoor activities that influenced my life. He announced in class that it was springtime and it was very good fun to take field glasses out to look at migrating birds in the trees on campus. He offered to meet at six o'clock the next morning to show us how to do that. Several of us showed up because we wanted to "apple polish" the teacher. At any rate, we admired him because he was the coach of the tennis team and a good athlete; we deduced that perhaps looking at birds was not a sissy thing to do.

After my first morning with him I was transformed into a dedicated birder. The only book we had in those days was the totally inadequate Reed's bird guide. But from then on, especially in springtime, I was up at five o'clock to go looking for migrating birds. One of my companions was a boarding student, Raymond O'Conner. I remember distinctly how we stood under a hermit thrush singing its heart out. We had even been told and knew enough to look at the bird when it sang its high register. Sure enough, it opened its bill but we could not hear the highest register song produced because it was above the frequency of human hearing.

My birding adventures at Andover led to a very satisfactory episode with my rather confident father. We were discussing woodcocks and how high they will fly. Of course we usually saw them very low at the edge of the woods or flying through the woods. But by chance I had heard them in courtship: this consists of the female calling on the ground and the male flying very high in the sky and then spiraling downward, singing a most beautiful song before eventually settling down alongside the female. But, as my father had grown up hunting to put food on the table, he assumed he knew all about woodcocks, and when I reported this he frowned.

"Impossible, absolutely impossible, a woodcock never flies higher than the trees," my father said.

"Well," I replied, "Come with me."

The next morning early we went to my woodcock's spot. Fortunately for my ego there was the male very high in the sky, coming down singing a beautiful song and landing in the grass very near us. My stern father actually used the sentence:

"Well, son you were right."

In my second year at Andover I became student librarian in the large new building. My best recollection of being on that job was being allowed to go into a special room and open the rare giant folio edition of Audubon's bird prints and turn each page after reading about each bird.

We had a dynamic biology teacher by the name of Larry Shields who had designed extensive laboratories for the biology class as well as a large new conservatory on the top floor of the science building. This conservatory influenced my entire life. There was a big skylight over a sizeable pool of running water filled with water lilies and aquatic animals. There was a raised rim around the pool, and all around it were large aquaria, cages, and garden areas in which plants could be nurtured.

In my second year, I spent so much time in the conservatory that Mr. Shields started taking me on field trips with the Advanced Biology class. He used his sports car with the top down to take us with nets and dredges to streams and ponds where we collected frogs, salamanders, fish and large aquatic insects. I'll never forget my first realization that the tadpole of a bullfrog has to live underwater for two years.

During my third year at Andover I was made curator of the conservatory, a paid position. I had already been catching snakes to bring into the conservatory and now I had charge of the whole facility. I was given a budget to send away for the usual lizards to display, like chameleons and horned lizards, as well as snakes, including a boa constrictor. The collection included two kinds of caddis flies (carpenters and masons), large predatory beetles and other specimens like dragonfly larvae.

I had to give up the interesting job as librarian, but part of being curator was going into the field to "catch frogs for the biology class." I had to fill the soil boxes in various parts of the room with plants, supply the aquaria with fish, fill the tanks with turtles, and stock the cages with reptiles. The great pool in the center also required constant attention. Of course, I thoroughly enjoyed every bit of this. Because my summer job waiting tables was such a contrast, it seemed unbelievable to be paid for doing the biology I enjoyed so much.

Now that I had a paid job at Andover, I was allowed to take over the family canoe, and the Shawsheen River came back into the picture. The old canoe house had fallen apart, and I now kept the canoe in the backyard of the Abbott Homestead. The sturdy canoe that had easily carried a family of five was not easy for one person to move, but somehow I managed. I remember hoisting the heavy canoe onto my back and then staggering along the road, under a railroad trestle, through the meadow that used to be the cornfield, and finally, to the edge of the river. Once on the river I could begin the pleasurable work of collecting. The trip to return the canoe to the Abbotts' backyard was about the same, only wetter.

But it was the canoe that made the collecting part of my job possible. Occasionally, Larry Shields would paddle the river with me, and whenever I had my collecting jars full he drove in his car to pick them up. I spent many weekends on the river collecting baby turtles and dredging for fish, baby eels and exotics. The Johnny darters—the fish that lacked air bladders and hopped along the bottom of the aquarium—were particularly popular in the conservatory, and helped attract students to take classes in Biology. Of course I filled the tanks with baby pickerel, perch and sunfish.

One day while I was working in the conservatory I learned of the excellent eyesight of a frog. The boss of the central pool was a bullfrog who must have been maximum size. He lived on a rock in the pool's center. One of our adult white mice had escaped and was running along the edge of the room. The frog left his rock, swam across the pool, hopped onto the floor, carefully approached the mouse and with a lightning-quick motion, snapped it up. He then returned to his rock. The frog had seen it from a distance of fifteen or twenty feet.

Our favorite animals in the conservatory were the snakes. I watched the birth of water snakes and realized that, reproductively, there were two kinds of snakes: those that gave birth to living young and those that laid eggs which had to hatch in the heat of the climate. I wasn't too fond of the unpleasant task of feeding live mice to the large black snakes, chicken snakes and milk snakes, but that was part of the job. Little did I know that in the future, snakes would do their part to help alleviate *The Financial Effect*, as well as influence my teaching style.

## *Bird Banding with Don Griffin*

At about that time, a young boarding student named Donald R. Griffin entered the tenth grade at Andover. I don't remember how we met, but because he was fascinated with biology, I'm sure he must have gravitated to the conservatory. In any case, we became fast and lifelong friends. He was a year younger than I, and until coming to Andover—indeed, until he began at Harvard years later—Don's education was, as he described it, "extraordinarily irregular" (Griffin 1985). He had received much of his schooling at home from tutors as well as his erudite father, and in his early quest to be a mammalogist, he had been guided by two curators at the Boston Museum of Natural History. As had I, Don had grown up exploring and investigating the fields, woods and waterways outside his home.

Sharing an interest in birds and their migratory patterns, we went to Larry Shields and told him we'd like to band birds. Mr. Shields was supportive, and because students weren't allowed to have bird-banding licenses, he said he would obtain one for the school. Then he told the maintenance staff we needed a building for a bird blind, and they located a shed that would do quite well. I remember watching the big truck bring the shed into the woods and deposit it there. With the help of the carpenter staff, we customized it, and soon had a slotted window for watching the birds.

Outside the window we set up drop traps: wire cages containing food and held open with a stick to which a long string was attached—when a bird entered we would pull the string and the cage would drop around the bird, safely trapping it. We would then carefully take out the bird, put a band on its leg, record the number, describe the bird, then release it. We sent our data to an agency in Washington, DC, and periodically received reports on where our birds had been spotted—one showed up in Texas. I remember one brave chickadee that continued to come back to the traps for the food. We figured we might have caught—or declined to catch—that same chickadee about 90 times!

Don was an exceptional researcher. We worked together later in college, throughout graduate school, and in the Harvard Fatigue Laboratory. We always kept in touch.

### *Looking Ahead, Graduation*

I was about to make a decision about entering college. As part of the process, I was sent to visit another cousin of my mother's, Sears Crowell, who was a graduate student in the Biological Laboratories at Harvard University. This is a magnificent, horseshoe-shaped building on Divinity Avenue, with an entrance guarded by two life-sized, one-horned, bronze rhinoceroses *(Rhinoceros unicornis).* Sears was friendly and showed me an experiment developed by the famous G.H. Parker. Using the skin of a small fish, Sears demonstrated how hormones can change chromatophores (cells that produce pigment), even when the nerves that supply them had been severed.

More importantly, because I was wondering whether in my senior year at Andover I should take specific courses before entering college; Sears introduced me to Professor John H. Welsh. He was a charming, friendly and informal person who sat on the steps of the Biology building while we discussed the courses at Andover and the freshman year at Harvard. Professor Welsh was later to become my undergraduate advisor and my PhD sponsor.

Back at Andover, I did well in Biology my senior year and became President of the Science Club. I invited Sears to speak to the club, and he graciously accepted. We stayed friends for life, and I often visited him in his laboratory at Harvard. Sears later became a professor at Indiana University.

Meanwhile there was another influence in my life, namely John Usher Monro. He lived up the street from me on Chestnut Street. We both were walking to the campus of Andover twice daily, a walk that took between twenty or thirty minutes. Thus we had plenty of opportunity for talk. He, too, was on a scholarship and had to work after school. He was a very bright person and a stimulating conversationalist. An interesting girl named Dorothy Foster lived in the house next door to me, and my acquaintanceship with John was enhanced when he courted Dorothy (whom he would later marry) by improvising jazz on the piano.

Her father was an editor of *Field and Stream* magazine and a fine artist, and as I was growing up, I enjoyed many conversations with Mr. Foster about wildlife. I remember that to my dismay, he declared war on snapping turtles "to save the baby ducks," and would spend quite a bit of time shooting the adult turtles when they were sunning on a dead tree. Nevertheless, he taught me to appreciate nature art because he painted many of the beautiful covers that graced the outdoor magazine.

I will never forget John's graduation from Andover in 1932. He was one year ahead of me and I attended his commencement exercises in a large, packed auditorium. Since the actual diplomas were given out in the chapel, at this part of commencement, only the awards were given out. The headmaster announced the first one, which was for the highest grades in Greek. John walked down the long corridor to receive his prize and went back to his seat. The next award was the highest overall grade in Latin. John arose from his seat in the back of the room to receive the prize and walked back. The next prize was in mathematics; once again, John walked down the long corridor and back again to his seat.

By this time a few people started laughing. The next prize was for the student who had contributed the most to student life. Once again, with a loud voice, the headmaster announced that it went to John Usher Monro. This time there was more laughter but the story wasn't over. There were at least two more prizes awarded and both went to John Usher Monro! I think one of them was for being the best orator. I'm not sure whether any other student received a prize that day.

John went on to Harvard, and encouraged me to follow (which in due course I did). While he was Editor of *The Harvard Crimson*, he would drop by my dorm room very late at night—as those were the hours the editors kept.

## *Exercising Body and Mind*

Besides trying to pass the difficult examinations, a major activity at Andover was the compulsory athletic program. This is worth discussing since in high schools in America today, physical education classes are electives; the only rigorous athletics are varsity sports like football and basketball, which are limited to the best athletes. This was far from the case at Andover. There was a doctor named Page in charge of the program.

As soon as a student arrived at Andover, he was required to exercise four afternoons a week. There were no classes between one o'clock and four o'clock, so Dr. Page expected us to be on the playing fields at that time. Page was a hard taskmaster. If you did not go downtown and buy soccer shoes on time for the first exercise, you would hear, "Certainly you know that we are not trying to pamper you here at Andover?"

A physical fitness test was required in order to graduate. The student had to run one-quarter mile, swim one length of the pool, climb a pole to the height of the rafters in the gymnasium, and complete the "belly grinds." This last consisted of hanging from a bar with your hands and touching your ankles to your hands ten times. Physical fitness at Andover was not to be taken lightly!

As I write these words from the vantage point of my nineties, I realize that the good health I have enjoyed must be partly due to that compulsory exercise at Andover. The literature of sports medicine now has many articles documenting how important exercise is for those from the ages of twelve to eighteen. By the end of four years, the pudgy, spoiled, shy boy who entered the school had become an aggressive captain of a soccer team and was able to run high hurdles competitively on the track team.

Fortunately, I did well enough on my national board exams that I could enter the college of my choice. I decided to join Sears Crowell and John Welsh at Harvard University.

# *Chapter Two*

## *Undergraduate Years at a Research University*
**1933–1937**

### Work-Study

THE FINANCIAL EFFECT was very powerful upon arriving in Cambridge. As had been the case at Phillips Andover, a scholarship student had to attain a B average to keep a full scholarship, and only one year was I able to attain that. Otherwise, I was on a work-study arrangement where I would earn room and tuition on one job, and board on another. The university was generous about providing jobs.

My assigned advisor, Jeffery Wyman, was a biophysicist and member of a prominent Boston family. He scheduled me to take five courses so I would get ahead rapidly. He based this heavier-than-normal course load on his assumption that as an Andover graduate, I would be able to handle anything. Unfortunately, he did not take the time to get to know me, or he would have realized that I was the kind of student who had to apply himself. Nor did he take into account that, unlike many of my classmates, I had to work two outside jobs. I quickly dropped one course.

The two jobs I held in my freshman year were at either end of the spectrum. One was waiting tables at the student union—I was quite used to this type of job because I had been a waiter at Fort Devens each of the previous four summers. At Harvard this job took six hours every other day and two Sundays per month. Students ate at tables served by wait staff (there was no cafeteria) and when the big dining room doors opened, there would be a waiter standing at the ready by each table. It was a busy job rushing to take orders and deliver food, yet a waiter could meet interesting people. One who sat regularly at my table was Franklin Delano Roosevelt, Jr., son of the new US President. He was a kind person and we chatted comfortably.

My second job was of a different nature. I was assigned to the biology stock room, where one of my tasks was to empty the crockery waste containers outside of each laboratory door. This was unfortunately before the age of plastic bags. The labs had many graduate students and post-docs working with Leigh Hoadley, the embryologist who would later become Director of Harvard's Biological Laboratories. Hoadley and his students spent all day working on chick embryos, discarding the rest of the egg and anything else into the crockery pots. Since I had five classes and another job, I often got to the crockery pots late at night or on Sunday morning. By this time, the contents had decomposed dreadfully, and I had to invent all kinds of systems to keep from losing my latest meal. For example, I could press a handkerchief firmly to my upper lip, but that imperfect strategy left me with only one hand to lift the heavy pots. I managed to survive, but it was not pleasant.

Working in the stock room provided more interesting experiences. One day a distinguished professor with a long beard reached across the counter and handed me a cloth bag. "This is a boa constrictor. I want you to euthanize and discard it," he said, then turned and walked out the door. Thus did I meet the famous G. H. Parker, who made his name doing experiments on the brain and spinal cord. In fact, the snake had been an experimental subject from which Parker had taken a very long nerve to study its impulse. I quickly and humanely carried out Parker's request.

## Other Activities

The university required that each student exercise five days a week the first year of residence, and I was only too happy to comply. I chose rowing on the Charles River in an eight-man boat, lasting through one season before the need for hours in the biological labs prevented my continuing that captivating sport. To fulfill the remaining requirement, I began taking life-saving classes and ended up with the highest certificate available at that time, which was called The Examiners Badge. This allowed me to teach swimming, which later turned out to be fortuitous *à propos The Financial Effect*.

In spite of the demands of jobs and studies, I could not quiet the voice inside that urged me to study birds. There was a very active bird club at Harvard, which of course I joined. It was in this group that I met David Davis, a graduate student in ornithology who not only became a close friend, but also profoundly influenced my career, as I'll soon explain.

The birding club was often fortunate enough to be joined by the renowned field ornithologist, Ludlow Griscom, who was a curator of ornithology at Harvard's Museum of Comparative Zoology (MCZ). Griscom was a genius who influenced the community by showing that birds can be precisely identified by observation without the need to collect specimens, saying "One need not shoot a bird to know what it is." On weekends we were up early, meeting at various automobiles to go into the field with him.

Despite the two jobs and many activities, I managed to stagger through my first year with a passing record.

## Advisor and Mentor, John H. Welsh

Like many undergraduates, I had to go through several advisors before I found the one who was right for me. After the first one had overloaded me with courses my first semester, I asked for a different advisor. I was then given Alfred Redfield, a Harvard professor who also held the post of senior biologist at the Woods Hole Oceanographic Institution, where he became famous for his many notable discoveries. He was Don Griffin's uncle, but they were worlds apart in charm and personality. The Redfield I knew was a remote scientist who resided in the clouds, avoided appointments, and could be scary and overpowering—to a student at least. He became a renowned oceanographer, but was not cut out to be an advisor—at least not mine! Sadly, the system is such that sometimes a professor takes on the role of adviser simply because it comes with the job.

Fortunately, I finally found an advisor who was not only a great scientist, but also a superb mentor. At the end of my difficult first year, I asked John Welsh—a biologist whom I greatly admired—if he would take me on as an advisee. John Henry Welsh, who liked to be called "Jack," agreed, and thus began a long and fruitful relationship. Jack

Welsh had earlier shifted his focus from the study of biological clocks to identifying the physiological functions of the neurotransmitter *acetylcholine* (ACh). He was one of the early investigators in that area.

Among my duties in Jack's laboratory was visiting South Boston drug stores to buy containers of live leeches (*Hirudo medicinalis*). Yes, leeches—the same blood-sucking parasites that swimmers in fresh water may find clinging to their skin. Leeches were used medicinally for thousands of years in bloodletting, a practice thought to remove "excess blood" or evil spirits. This antiquated practice was abandoned in the 19th century, but apparently in the mid-1930s, some Boston doctors were still using leeches to cure their patients' ailments. Today, interestingly, leeches are again being used medically, but in a different way—it has been found that in some cases they can help restore damaged or surgically grafted tissue.

In his laboratory, Jack exposed strips of leach muscle to ACh that had been diluted up to one hundred thousand times ($10^{-10}$). He found that even this low concentration was strong enough to make the muscle contract. Thus, his bioassays established that ACh, which exists in most organs, is far more powerful than had been previously suspected. Because of his work, ACh is now known to be an important cardiac hormone. Over his career, Welsh published almost 30 papers on the distribution and mechanism of action of ACh.

Later, in the early 1950s, Welsh would become known for the important discovery that the substance *serotonin* (5-HT), previously known as a powerful vasoconstrictor, was actually an essential neurotransmitter. Today there are at least 60 identified neurotransmitters; 5-HT was the third to be discovered, and the second was ACh.

Jack was a wonderful mentor who cared deeply about his students. Over the years, I found we shared many experiences, ranging from fascination with reptiles to having to work at outside jobs in order to pay tuition. I was indeed fortunate to be one of John Welsh's students. He remained my advisor throughout the rest of my undergraduate years, and later he was my sponsor for my Masters and PhD degrees.

## Ornithology

In my second year at Harvard, things got a bit easier—I was made a student librarian in the biology library and could sometimes study at work. I quit waiting tables but continued my job in the Biological Laboratories' stock room.

I signed up for a course called The Biology of Birds, given by Glover Allen. Up to this point I had been tempted to become a typical ornithologist or perhaps a taxonomist of reptiles. I could imagine spending a lifetime going to picturesque marshes where there were twisted streams of fog and moisture, all to identify migrating and resident birds.

But one episode during Prof. Allen's course changed all that. He described an experiment to determine whether birds could distinguish colors. It was a simple experiment in which a prism broke up light into primary colors; this spectrum of light illuminated grain offered to the chickens, and researchers watched to see which color of grain the birds selected. To analyze whether the light affected how the chickens saw the corn, the researchers counted the number of grains of corn eaten from each band of colors. After hearing about this experiment, I became so intrigued by this combination of experimental biology and ornithology that I decided that this was the kind of thing I wanted to spend my life on.

Interestingly, my good friend and fellow birder, David Davis, had also advised me not to become an ornithologist, but for different reasons. First, he said, the ornithology arena was far too crowded. And second, he pointed out, it was dominated by many who were able to finance their own field work—these capable ornithologists could outfit their own boats and study birds around the world. "It's a millionaire's hobby," he pronounced.

The knowledge that ornithology might not prove a suitable career for me did nothing whatsoever to dampen my long-held pleasure in watching and identifying birds. I continued my forays with the bird club and in the process made another life-long friend in Garret Eddy, a first-year biology student and ardent birdwatcher. Garret eventually took me to the prestigious Nuttall Ornithological Club, where, with its long history of influential and important members, I could rub shoulders with such greats as Ludlow Griscom and Roger Tory Peterson.

## Snakes – Biology Provides an Income

At about that time, *The Financial Effect* became more demanding. With my two work-study jobs, I was not concerned with paying tuition, but I needed cash! The student employment office was a friendly and helpful place, and offered many odd jobs. At first they sent me to various houses in Cambridge to clean fireplaces and chimneys.

Fortunately, another opportunity to earn money soon arose that was directly related to my interests and experiences in biology—in particular, my experience with snakes. I had been volunteering at Harvard's Phillips Brooks House, a center for charity and philanthropy, and my particular specialty had been to lecture about snakes. Thanks to my role at Andover's conservatory, I knew snake biology, was comfortable catching and handling snakes and had learned to order harmless serpents though the mail.

One of my first assignments as a volunteer lecturer was just outside of town at Camp Blue Hills, a base of the new Civilian Conservation Corps (CCC). President Roosevelt had set up the CCC under the New Deal legislation designed to help the nation recover from the economic devastation of the Great Depression. I was pleased to lecture to the participants and to learn about their projects, which included reforestation, erosion control, and construction of bridges and dams—to name just a few. For my lecture, I had borrowed some snakes from Phillips Andover and soon found that my living examples elicited avid interest from the audience about the biology of reptiles.

When the student employment office heard that I had been giving "snake lectures," they freed me from cleaning chimneys and began to get me jobs as a lecturer, sticking with my unusual, but popular, topic.

Since it was not practical to keep borrowing snakes from Phillips Andover (nor was it appreciated by the biology teacher!), I ordered a bull snake and a kingsnake from a supply house in Iowa. I also caught a water snake back on the Shawsheen River in Andover. These three snakes would be the principal actors on the stage at my lectures.

Soon I was traveling with my snakes to various institutions around Boston, mostly to church and Boy Scout groups, but also to Northeastern University. I received a fee for giving these lectures, and did not always have to use it for tuition. I never quite dared to tell my Radcliff date, however, that the only ready cash I had to take her ballroom dancing was from snakes.

**With one of my "Dancing Money" Snakes (Bull snake from Iowa).**

Probably the main reason the lectures were successful was because my water snake quickly learned to eat dead goldfish from my fingers in front of the audience. All three of my snakes were very docile and could even be handled by members of the audience after the lecture. The other two snakes were spectacular specimens—large and brightly colored. The bull snake is one of the largest snakes in North America (often as long as eight feet, although the record for largest goes to the eastern indigo snake, at more than nine feet). Kingsnakes come in two dramatic patterns, "chain" and speckled; I had the chain variety, named so after the yellow bands crossing its black body. Audiences were always fascinated to learn that kingsnakes kill and eat poisonous snakes. I enjoyed having these particular snakes because they occurred in New England, being found as far north as New Jersey.

When they weren't on stage, the snakes lived in the dormitory with me. I had a scholarship room with three roommates on the top floor of Winthrop House. It consisted of four small bedrooms with a large living room. There was even a small spare room where I could keep the snakes, as well as any other biology specimens.

Each of my demonstration snakes deserves a special story. The water snake always thrilled audiences when he would crawl across the table and take the goldfish from my fingers—this was to show the audience how a snake could swallow its prey whole. The large bull snake was a favorite of the newspaper reporters who liked to take pictures of this strange undergraduate student who could lecture on snakes.

The kingsnake represents a totally different story. This magnificent snake was a favorite of the other students in Winthrop House. These young men would borrow the snake, tie a ribbon around its neck finished off with a bow, and then show off their "bravery" to their girlfriends with this gentle and harmless—but big and flamboyant—snake. Unfortunately, one time when a student borrowed the kingsnake, he did not put the cover on the cage tightly and the creature escaped. I knew that the room in our dorm where I kept the snakes was connected through the water pipes with the top floor of Winthrop House, which was the Headmaster's residence. I knew as well that his daughter had her room on the other side of the partition where the pipes went through from our apartment.

I went through a long week of worry and dismay, fearing that the vividly-colored snake would crawl through the pipes and end up in the bathtub of the Headmaster's daughter—and that she might have a shock or faint upon seeing it. What a relief it was when one evening I came home to find the kingsnake stretched across the doorway; I picked him up and put him back in the cage, he at least, was no worse for wear.

I know that giving snake lectures was outstanding training for my future career as a teacher. I quickly lost fear of any large audience, and because of excellent training I'd received at Phillips Andover, knew from the beginning to project my voice and select a person in the back row to direct the lecture to. I attribute my charming, agreeable and helpful snakes as being the clue to a key point: interest your audience!

When Professor Welsh learned I was giving lectures on snakes, I discovered that he shared my interest in herpetology. He gave me a copy of an article he had written years earlier, for which he had caught his own rattlesnakes in western Massachusetts. Jack had also collected their venom, which led to his life-long fascination with animal poisons. Over several decades, he conducted a series of studies to isolate and characterize transmitter agents in the toxins of many animals ranging from snakes to insects and spiders.

## Sports – A Route to Income and Outdoor Adventure

Each summer I worked at jobs off campus. My first year I worked as a camp counselor, but that had been an ill-fated venture because the owner was a scoundrel who reasoned that since his camp was on a lake he didn't need to pay his employees very much. To my parents' disgust, for that entire summer he'd paid me only $50.00.

**Camp counselor, Freshman summer at Harvard.**

My second summer I was not taken advantage of—quite the contrary. I was hired by Clark Macomber to be a companion to his son George who had asthma. To keep George away from the city air, the family lived in the upstate Massachusetts town of Winchendon. Clark, the contractor who had built the Boston Post Office, commuted 80 miles to the city. I was delighted to spend the summer in Winchendon, which was close to the New Hampshire border and not far from Mount Greylock, the highest point in the state.

I basically ran a summer camp for the Macomber's children and some of their friends, a number of boys of equal age. I gave swimming lessons—thanks to my advanced life saving badge—and Clark bought a station wagon and handed me the keys so I could take the children to all sorts of fun places.

The Macombers were great people. Janet, the mom, was an outdoor type and as we took the children on hikes in the mountains, we became good friends. Clark had been an All-American football champion, and although he had become a politically powerful man, he was a kind and warm person.

For two summers I enjoyed the splendid company and position provided by Clark Macomber and his two outstanding children, Gail and George. We all loved to go hiking, and New Hampshire's rugged White Mountains provided endless opportunities. The White Mountains have 48 peaks over 4,000 feet, as well as the highest mountain in the Northeastern United States: Mount Washington, at

6,288 feet. We often went to one of the closest and most picturesque, Mt. Chocorua, which rises to 3,490 feet.

Eventually, the two children became outstanding skiers: George was an internationally ranked competitor who won a place on the men's US Olympic team in 1952, while Gail became the Women's North American downhill champion.

After the first successful summer with the Macombers, a big event in my life occurred: that Christmas, the Macombers gave me a pair of downhill skis, boots and poles. I quickly went off with friends to Mt. Washington and learned to ski.

## Skiing and Winter Camping

Fortunately I had plenty of motivation to do a winter sport. The winter before, four of us had visited a former lumber camp on the backside of Wildcat Mountain, not far from Mt. Washington. There we had cooked our own food and been taught (by the owner of the camp) to make homemade toe-strap skis by fastening a loop of inner tube rubber from the back of the ankle and over the toe.

We had climbed Wildcat Mountain and followed the tracks of a black bear along the ridge. When we began our descent, I'm sure the sight of us crashing downhill on our makeshift toe-strap skis appeared ludicrous to any real skiers who saw us. But every morning we had awakened to see the sun shining on Carter Dome, one of the most beautiful mountains in Pinkham Notch. The excitement of being on Wildcat Mountain in winter convinced me that I should use every excuse to spend as much time outdoors as possible the rest of my life. All of us had vowed to return frequently to the White Mountains.

And now, because of a gift from the Macombers, I had real skis and the chance to try them out on Mt. Washington, even skiing down the steep Carriage Road that runs along a ridge on the east side. I will admit that I learned rapidly and began to have fun on skis, especially on Mt. Washington's two trails, the Fire Trail and the Sherburne Trail.

From then on, heading to Mt. Washington with my skis over both Christmas and Easter breaks became my vacation of choice. My parents weren't pleased about this, although I always spent at least one of my vacation days with them. I didn't have a car, but I could always find someone who had one and was eager to go. Once, we even made the 150-mile wintertime drive in an open Model A! As I sift over the names of my frequent skiing companions, the most important ones who come to mind were Garret Eddy, my biology and birding pal, and "Trader" Horn, who had learned to ski in the mountains of his home state, Colorado.

A question may come to mind: How could a college student struggling under *The Financial Effect* find the finances to go to ski areas with their ski tows and expensive motels? The answer was a hut on Mt. Washington; it belonged to the Harvard Mountaineering Club, and even though I couldn't afford to officially join the club, they gave me permission to use their hut. Several of us in the same predicament—with a love of skiing and a flat pocketbook—learned to pool resources, even for a weekend. We left on Friday afternoon and drove to Pinkham Notch. With knapsacks and skis, we climbed the Fire Trail to the mountaineering hut, which was in the woods just below the entrance to Tuckerman's Ravine. Snow was always so deep that it was hard to plod through to the cabin, but we always felt it was worth it. The toilet consisted of facing an evergreen tree or putting your back against it.

There we spent some delightful skiing time for three years, especially during Easter Vacation. Tuckerman's is known for having the best spring skiing; it was (and still is) such a popular place that later in the day there might be 1,000 skiers in action there. Staying at our primitive hut meant beautiful days of freedom and being the first people each morning in Tuckerman's Ravine.

I remember one memorable Sunday on the mountain before Tuckerman's was open for the spring season. Under a beautiful clear blue sky, we skied the entire Sherburne Trail from top to bottom five times. The last run down and off the mountain, carrying knapsacks full of sleeping gear and empty cans made it especially exciting.

In May, the alpine skiers arrived. I'll never forget one expert skier from Austria, a woman who dared to wear a long flowing blue skirt. She traversed broad reaches of the great snowy bowl, her garments fluttering like an artist's brush traveling back and forth across an expanse of white paper.

These experiences on Mt. Washington proved valuable to me later in life, particularly during WWII. I even credit my knowledge of winter camping with saving my life during that war because it played such a strong role in determining where and how I served; likewise, my familiarity with carrying food up the mountain through snow would be especially useful when I helped design emergency rations and winter clothing for our troops.

I find it interesting to consider the many ways in which *The Financial Effect* changed my life for the better. In this particular case, I can trace my need to earn money to working for the Macombers, who became good friends and gave me skis, which led me into winter adventures and camping, which later opened the door to the unusual and interesting wartime research that probably saved my life, and definitely shaped the rest of my career.

## Bats

During my junior and senior years there was another outdoor winter activity, which was even more enjoyable because it involved collecting scientific data. I joined a very active group of biologists led by Donald Griffin, my close friend from Andover with whom I had run a bird banding station. Don's interest in the outdoors went beyond collecting and identifying birds and mammals. In spite of his youth, Don was already a full-fledged scientist who was primarily interested in experimental biology. While still an undergraduate, he had received funding for an audacious research project: to study the habits and life histories of New England bats—or in his words, "to band all the cave bats in New England." It was typical of Don to think on such a large scale, and I always enjoyed collaborating with him on his inventive and original research.

This study was indeed science on the far horizon. Don made a card catalogue of all caves as well as active and abandoned mines in New England. His goal was to visit each of these locales and to collect all hibernating bats, then band and release them. He knew this study would span years, and that by recapturing our banded bats, he would discover the answers to many questions about their migration, hibernation and method of flying. I was among the group Don assembled to visit all these caves and mines, as was my friend Garrett Eddy, and many others. Two of these were Peg Robinson from

Wellesley College and her brother Douglas, who carried on the tradition and banner of outdoor research with his own studies, which included muscle hemoglobin in harbor seals. Another bat bander was Jane Bridgeman, an avid outdoorswoman who was also Don's then-fiancé.

One unlikely comrade in these outdoor adventures was Robert Galambos, a brilliant student who admitted he was not an outdoorsman and did not like caves. He got involved, however, because he wanted to keep some of the bats to do his own auditory research with them back in the laboratory. Robert wanted to catch his own bats so he would not be indebted to Don, but those two eventually did team up to conduct ground-breaking studies of bats' amazing auditory abilities. Later, as a neuroscientist, Galambos became renowned for his pioneering work on brain responses to auditory stimuli.

Don also invented a cage for gathering bats and for containing them while banding or reading bands. He gave us each a cage, and depending on the cave and access to the bats, it would sometimes feel a little like picking blueberries: pluck a hibernating bat and put it in the cage.

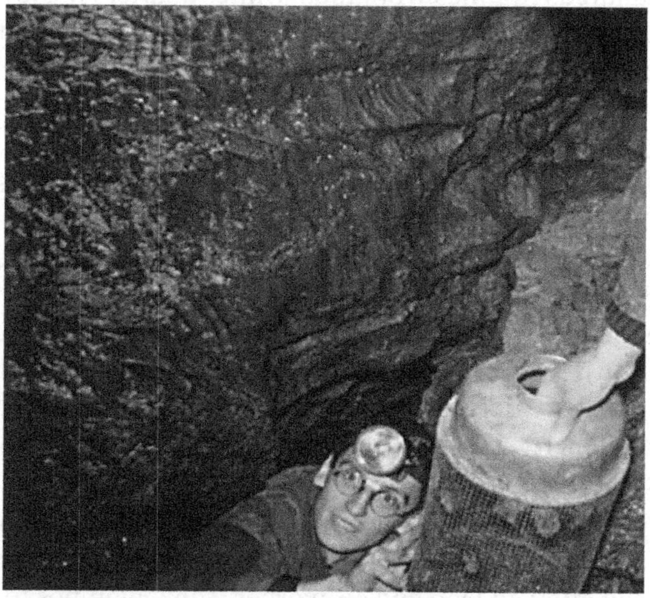

**Don Griffin emerging from a crevasse with bats in a cage. (Photo courtesy of Griffin Family.)**

It was delicate work holding a live bat and attaching a miniscule aluminum band to its hind leg—too delicate for us to wear gloves. Consequently, we were bitten by hundreds of bats. The bite seldom drew blood, especially with the smaller species. It was quite different, however, with another bat, *Eptesicus fuscus*, commonly known as the Big Brown Bat. These bats are five inches long with a twelve-inch wingspan. As we banded them, they sometimes managed to twist their heads and bite, and they always drew blood; hence, we nicknamed them "wolf bats."

Years later it was discovered that bats are symptomless transmitters of rabies; why then did none of us come down with that disease? It was not until the 1960s that rabies was first found in bats anywhere in this country, thus it is likely that in New England in the 1930s and 40s, there was no endemic rabies in bats. That would explain why, despite the number of times we bat banders were bitten, none of us got rabies.

Don's research had a simple secondary thesis: Where do bats go in the summer when they leave the caves in which they have spent the winter? Perhaps to far Northern Canada, or maybe a more westerly direction? His study of recaptured banded bats showed that some bats migrate hundreds of miles from their winter quarters in New England to a range further north and east where they spend summers in trees and buildings. One of the most astonishing findings over the years was the unusual longevity of bats. While similarly sized creatures like the laboratory mouse and wild house mouse live for only two to four years, Don's banding program showed that bats are much longer lived, sometimes even reaching 32 years of age.

Years later, in 1946, at the end of Don's 14-year study, he set down what he'd learned about bats in a superb first-person article, "Mystery Mammals of the Twilight," which appeared in *National Geographic* (Griffin 1946). In it, he corrected popular misconceptions about these harmless and beneficial animals—they are not blind and do not get tangled in women's hair—and he unraveled some of the mysteries of their hibernation, migration, and reproduction.

Most important, however, was that he described bats' extraordinary ability to avoid obstacles while in flight. Developments in physics by that time had allowed Don and co-worker Bob Galambos to detect the supersonic cries made by bats; these sounds return to their brain through their ears as echoes, allowing them to detect and avoid objects, navigating around and between them easily. Griffin invented the word echolocation to describe "the location of objects by means of echoes." Echolocation is also how bats catch the flying and darting insects upon which they exist.

In the article, Griffin described bat-banding adventures in "innumerable dank caves, abandoned mines, and crannies where only a bat would feel at home." Although more than thirty people had participated in the bat banding effort, Don acknowledged a core group of "outstanding…master bat banders," and listed eight of us by name. I was proud to be among them, along with Garrett Eddy and Doug Robinson.

**Donald Griffin, *the* master bat bander. (Photo courtesy of Griffin Family.)**

The number of bats banded when he wrote that article was more than 14,000. Most of these were cave bats; these are the hibernators and stay put, while tree bats migrate south for the winter, as do many birds. It was not unusual to find hundreds to thousands of the gregarious cave bats hibernating in one place. But it was difficult to band any of the three kinds of tree bats; one reason is that they do not congregate in colonies as do the cave bats, and it is not easy to find enough specimens to do any "worthwhile banding," as Don explained in the article. One of the tree bats, *Lasiurus borealis*, the red bat, is especially attractive since it has a coat like a red fox. Of the three species of cave bats, the smallest is the little brown bat, which has the scientific name *Myotis lucifugus*.

I've always found the Latin naming system fascinating and will endeavor to explain a little about it with the following example. A major group of similar animals are collected together under the umbrella called a Genus. *Myotis*, for example, is the genus for mouse-eared bats. (The genus part of the scientific name always begins with a capital letter.) Within the genus, a particular kind of animal is called a species: *Myotis lucifugus* is one of 90 species of mouse-eared bats. (The species part of the scientific name always begins with a lower case letter, and the entire name is italicized.)

You can imagine the thinking of taxonomists when they learn that they have a new species and they must find the appropriate Latin term. Continuing with our example above, the Latin word *lucifugus* is translated as "light-fleer." There are two other unrelated species that have *lucifugus* as part of their scientific name. Apparently, taxonomists named these three organisms "*lucifugus*" because two of them are blind and the third one lives in caves. Their complete scientific names are *Termes lucifugus*, *Cambarus lucifugus*, and *Myotis lucifugus*. The first animal that flees the light is a termite, the second is a blind crayfish, and the third is our small brown bat. This is one of the most beautiful examples of the naming system: these three unrelated animals are in three different genera, but the species name uses the same word because it is so descriptive.

My experiences with Don in field biology inspired me to carry on in a similar fashion, and bats continued to figure in my life, and, as you'll see later, even helped me get into graduate school.

## White Water and the Old Canoe

One of my three college roommates, William Bersen, had a group of very energetic friends from Milwaukee, most of whom went to Dartmouth College. Two of his friends, Harry Espinchard and Fritz Myer were in graduate school there, but because they wanted to use the libraries at Harvard, they slept on our living room floor. They lived with us like that for most of my third year.

Fritz Myer had been president of the world-famous Dartmouth Outing Club the year before. During a meal one spring evening, Fritz bemoaned the fact that this was the first year he was not traveling down forty miles of whitewater on the White River in New Hampshire and Vermont.

"For four years I led the canoe club down those miles of white water," he said, mournfully," and this year I have to miss it because I don't have a canoe."

I perked up. "But I have a canoe."

I had not in fact seen the canoe since I graduated from Andover three years earlier. I think the Abbotts used it occasionally, but mostly it remained in the back yard of the Abbott Homestead.

Fritz's eyes practically glazed as he asked, "Where is it?"

"Well, it's not very near here," I told him, "it's in the town of Andover about thirty miles away."

"Eddie," Fritz asked hopefully, "will you go down the White River with me?"

I had my father's inclination for adventure so in spite of course work, I said, "Sure."

Fritz had a Model A touring car and before I knew it, we were in that car headed for the canoe. We fastened it on the top of the Model A and headed for Dartmouth College. I had no idea what I was getting into. I assumed we would have a most pleasant time "shooting" the rapids and enjoying the ride. As it turned out, the adventure was nothing like that. I had never heard of the words "steerage way." What that means is that to avoid the many rocks in a whitewater river, you have to be paddling faster than the water is running. Otherwise, the water takes you against the rocks.

Eventually we arrived at the White River. It was spring, and I could see that about 50% of the water's surface was covered with white foam. I had seen rivers like that before but only from a distance. This was to be different from my typical canoe trips on the Shawsheen.

The next day, a canoe club with about fifteen canoes assembled on the water forty miles upstream. It turned out that Fritz was a celebrity and received many congratulations. He had the reputation of not "dumping," because for many of the years he was in the club, he was the only one who did not tip over during the forty-mile stretch. We left the shore and when the first rocks appeared, I heard the shouted command, "Paddle, Eddie," and I paddled as fast as I could. Thus we obtained steerage way and avoided the rocks. It went on like this for half of the day.

We had left in the afternoon and we spent the night in a farmhouse. Naturally, as a plodding student at a university, I was not in condition for such strenuous activity and my arms ached. That night Fritz briefed me for the next day. He said, "We go over the dam tomorrow and you have to be very careful. Last year two members of the canoe club drowned because they did not hold onto their canoe after they shot the dam. The water was so full of foam they could not swim. Be very careful, Eddie."

Of course, I didn't know exactly what he meant by being careful, but I guessed that holding onto the canoe was the main thing to do if we tipped over. Away we went, and although I had an impression of flying through the air, because of Fritz's expertise, we did not tip over at the dam. I don't know how many of the other canoes did tip over, but everyone survived. We did take on plenty of water over the side, and needed several stops along the bank that twenty-mile morning to tip water out of the canoe. We finally reached the Connecticut River, which was a very exciting moment for me. Then we drove back to Andover and put the canoe in its place behind the Abbott Homestead.

That spring my family moved out of Andover and my father took a preaching appointment at a small parsonage outside of Worcester, Massachusetts. What happened to the canoe after that, I never heard.

## The Gaspé Peninsula

The following summer, I wanted to make an expedition to conduct some field work on the far horizons. In this case I wanted to collect insects from the Gaspé Peninsula of southern Quebec, Canada.

I had been motivated by a professor at Harvard, Phil Darlington, who was also Assistant Curator of Insects at the Museum of Comparative Zoology. Phil was a taxonomist interested especially in insects and reptiles; he was such an inspiring teacher that many of us became interested in biogeography, the study of the distribution, past and present, of plants, animals and insects over the earth. (Later Phil wrote the definitive text on the subject, *Biogeography*.) Phil had gone on collecting expeditions in the West Indies and Colombia, and told me he needed insects from an area here in the Northeast at the northernmost end of the Appalachian range—the Chic-Choc Mountains—because that area had not been glaciated at the time of the last ice sheet.

The Gaspé Peninsula of Quebec extends into the Gulf of Saint Lawrence; its northern shore borders the St. Lawrence River, and its southern edge borders New Brunswick and a small section of northern

Maine. The area's official name, *Gaspésie* is derived from a word used by the early native inhabitants that means "land's end."

It was the end of my second summer working for the Macombers. Clark was enthusiastic about the expedition I'd planned and wanted to help. He loaned me a truck and introduced me to two young men who wanted to go with me.

It took us nearly 20 hours to drive the more than 700 miles from Winchendon to Gaspé. I knew there was a road that followed the coast around the peninsula because my parents had toured that area many years ago. My companions and I camped beside the truck, and slept in deserted barns when it rained. I saw the famous *Rocher Percé*—the massive rock that lies off the coast near the town of Percé; it is said to predate North America. I watched seabirds and took pictures of gannets, which are big white diving birds.

Then we turned inland. At that time, there were no trails into the rugged interior, so we hiked through a trackless northern forest, clambering over deadfall, looking for insects for Phil. We were excited to find the mountains inhabited by woodland caribou, which I had not seen before. Insects, however, were scarce and I ended up with fewer than 20 in my vials.

I never did learn whether any of the insects I brought back from that trip were useful to Phil or not. It happened that several years later, as a graduate student, I would collect more specimens for Phil on treks in the West Indies, and I know *those* were useful.

## Doc Edgerton Shoots My Snake

A delightful experience occurred when in my senior year, I wanted to learn about poisonous snakes, and I purchased a rattlesnake to keep in a glass case. (I didn't know it then, but later on, in the hills of Iowa, I would capture my own rattlesnakes for collecting venom and for teaching.) John Welsh introduced me to Harold Edgerton, who had a lab at MIT, and was becoming famous for developing the electronic strobe light and the ultra-high-speed camera. One of his best-known photographs was of a drop of milk falling into a saucer, with all the splatter being caught by his camera at a speed of 120$^{th}$ of a second. He also took the first photos that "stopped" the wings of flying hummingbirds, and later, with Don Griffin, took the incredible photos for *National Geographic* which revealed that bats in flight make about 15 wing strokes each second.

When I mentioned the rattlesnake's strike to him, Dr. Edgerton, or "Doc" as he was known, said he would like to stop that action with his high-speed camera. He knew that, as the rattlesnake strikes it opens its mouth, pulls back a membrane, and protrudes the fangs, ready to inject poison.

I took my rattlesnake in a glass case along with several mice to Doc's laboratory. His high-speed camera could shoot up to 15,000 frames per second; Doc told me that when he pressed the trigger on the camera about a hundred feet of film would fire rapidly and he wouldn't be able to stop it. Thus, I was given the difficult job of saying when the snake was going to strike.

The snake was very interested in the lunch we'd provided it, and stared at the mouse, obviously deciding how and when to strike. Three times I thought the strike would occur, but the snake did not cooperate. Twice when I gave the alert, the film ran through the camera but the snake didn't strike, and the film was wasted. The third time I said "Now!" the snake struck and drew back again so quickly that the film had not even started running. Doc was very good natured about all this and took out another camera to photograph the rattlesnake as it ate its lunch, which had died instantly upon being bitten.

## More Snakes

During my senior undergraduate year, I had another experience with a fine scientist who liked to do research with snakes, namely Hermann Rahn. He came to the department as a post-doctoral fellow in 1936, and I was often in his research laboratory where he studied the endocrine glands of garter snakes.

When I found out he needed more garter snakes to continue his work, I told him I knew exactly where there was a hibernating den in Andover, and offered to take him there. We drove in his car about 25 miles to the den, and since I was very impressed to be traveling with a post doc, I was timidly hoping that my past experience really would bear fruit (not to mention snakes). Hermann didn't know that I held my breath as we approached the site. Sure enough, to my delight and relief, near the den we were able to pick up a whole bagful of garter snakes.

That was the beginning of a long and mutually beneficial friendship and collaboration; when I later became a graduate student, Hermann's interest in reptiles would expand my horizons and influence my research. Hermann went on to do great things, and when he was elected President of The American Physiological Society, he entertained an audience with the story of the time I took him to capture a den of garter snakes.

A few years later, Hermann would send Doc another rattlesnake, this time from Wyoming, to use his stroboscope to measure the frequency of the rattlesnake's tail as the snake gave its signature warning. I know they were successful, though I'm not sure what the frequency of the rattle was.

During the time I was lecturing on and studying snakes, I received considerable encouragement from the Department of Herpetology in the MCZ. The museum's curator of reptiles, Dr. Loveridge, gave me two dried Bushmaster heads, which had come from South America. The Bushmaster is the largest pit viper in the world, with the record being 14 feet. Pit vipers are venomous snakes that can detect the body heat of their prey with heat-sensitive hollows or "pits" that lie between the eye and nostril on each side of the head. Unlike other pit vipers that are live-bearers (for example the rattlesnake) the Bushmaster lays eggs.

These large poisonous snakes are somewhat unique amongst the snake group in that they are very aggressive. Most snakes around the globe, including rattlesnakes, will make every effort to get away from you and go in the opposite direction of your path. But three snakes worldwide stand out for their aggressive temperament, the Bushmaster, the Mamba, and the non-poisonous black snake of the United States. In the latter case, if you venture onto its territory or corner it, a black snake may attack and bite your legs, although a thin pair of trousers will protect you. The deadly Bushmaster of South America will actually chase a person and attack. Alexander Wetmore, the outstanding ornithologist and Director of the Museum of Natural History in New York City, was once bitten by a Bushmaster and nearly lost his life.

Knowing this history, I was excited to be given the big Bushmaster heads so I could prepare the skulls down to the bone and expose the poisonous fangs. When I was home on vacation, I boiled the heads in soap and water on the stove and began to pull off the chunks of dried flesh. Of course, I managed to prick my finger on one of the fangs but fortunately was correct in assuming that boiling in soap and water would counteract any poison remaining in the fang. I was able to make completely clean bony skulls showing the teeth pointed back within the mouth.

To my surprise and that of the museum, the mouth of one of the snakes contained four fangs instead of just two. Normally there is only one pair of functional fangs; in this case it's likely that the old fangs were being replaced by fresh ones but the snake had been killed before the first set dropped off. At any rate, the skulls made a good display for my lectures.

I learned that Bushmasters are common in some locales in South America. For example, when an abandoned coffee plantation on Trinidad Island was being turned into the Asa Right Biological Station, workers trimming the bushes around the old coffee plants had to kill fifty Bushmasters.

At a much later time in my life, my knowledge of this animal affected my behavior. My wife and I had traveled to Guatemala and wanted to visit the Maya ruins. As we got on the bus in Belize to drive to the ruins, the driver told us: "Ladies and gentlemen, we have to take longer in our drive to the ruins, because there's been an upset: the main highway is now blocked by men carrying guns, sticks, and stones. They will not allow any vehicles to pass and we must go a different way." (This was the driver's way of telling us that a revolution had erupted.)

The roundabout way was entirely through rain forest, which we enjoyed because some tropical birds flew across the road as we progressed. At the end of the visit to the ruins, we had to drive back the same way at night. There were no facilities or gas stations or restaurants on the entire trip through the rain forest, and therefore, the bus had to allow people to wander into the forest in lieu of a toilet. It was totally dark and there were no flashlights. I knew too much; I had measured 3-inch fangs in Bushmaster skulls. I did not feel free to remind the other tourists about snakes, but no matter how uncomfortable I became, I would not venture into that rain forest at night.

In 1937 I laboriously managed to graduate from Harvard University with a BA in Biology. I was never a good student, and the jobs I worked at during all four years certainly affected my grades.

Again *The Financial Effect* came to the fore; I was determined to go to graduate school, but there was no source of tuition funds. In those days and even up to the period after WWII, there were few scholarships and fellowships for graduate students. To build up my bank account and fill in the time before graduate school, I felt that teaching in a preparatory school would be satisfactory since there seemed to be numerous positions available. The position I accepted was at Salisbury School in Connecticut.

# Chapter Three

## First Teaching Position, Salisbury School

**1937–1938**

THE YEAR I spent teaching at the Salisbury School in Connecticut was one of the happiest of my life. Not only was the school's location on the southern edge of the Berkshire Mountains absolutely charming, but also its faculty was so alert and interesting that this recent college graduate found stimulus and intellectual challenge with almost every conversation. The Headmaster, Emerson Quaile and his wife Charlotte were wonderful people who mentored me and brought me into their lives as friends.

I was actually hired to teach a group of younger boys, including the Quailes' son, George. These young men were not old enough to be enrolled in the formal prep school classes, but there was no school in the area deemed suitable for boys their ages. The parents of the other six boys took turns driving them up to Salisbury School, where I taught special science classes. I filled several classrooms with animals, taught daily lessons, took the boys on field trips, and assigned appropriate homework. Salisbury School was run very much in the English tradition, with instructors wearing academic gowns as we taught.

Aside from teaching, there was also the excitement of interacting with faculty from the neighboring prep schools, Hotchkiss, Berkshire and Kent, and they swept me up into their active book clubs. Fresh from being a senior work-study student at college, I soon learned that older people were friendly and surprisingly human.

I also received my first paycheck for teaching. I used it to buy two hand-colored bird engravings, which I still own and admire. Salisbury School had hosted an art exhibit soon after I arrived, and these beautiful engravings of a pair of European Kestrels immediately caught my eye. One is dated 1826, and signed by P.J. Selby (1788-1867), the British ornithologist and contemporary of John James Audubon. The other engraving is obviously of its mate, but doesn't have the signature. I couldn't resist these prints because, aside from my love of birds, the art classes I'd taken from Professor Clifford Pope at Harvard had turned me into a dedicated art advocate with plans to attend all important retrospectives as they traveled through the United States.

The village of Salisbury sits at the far northwestern corner of Connecticut in the tri-state area close to both Massachusetts (to the north) and New York (just west). Natural beauty abounds, with steep hills and forests. The Berkshires are modest mountains; the state's highest point, 2,380 feet, is the south slope of Mt. Frissell, near Salisbury.

Behind Salisbury School is a low mountain crested with a rocky outcropping that was covered with tundra-like vegetation. There were great views from its bald top. Down the hill on the other side of the school are the Twin Lakes, Washinee and Washining, beneath which runs an extensive system of limestone caves. In the early morning, it was actually possible to dash down to the lakes before classes and swim. There was also a ski jump on the front lawn, and between the two directions in which to take exercise, as well as skiing in the winter, we all seemed to stay physically fit.

Perhaps best of all was that I could continue doing field biology in this rich natural environment.

## More Bat Research

There were at least two caves in the area with hibernating bats, the Twin Lakes Cave beneath Lake Washinee and the Indian Oven Cave on a hillside not far into NY.

I'd heard about the cave below the lakes from Donald Griffin even before I'd moved to Salisbury. Throughout the cave, narrow limestone tunnels connect a series of large caverns, which I never did see. The one time I was in that cave was at the request of Don, who had earlier banded bats there and wanted me to read the bands on all the bats I could find in the cave's narrow tunnels. It surprised me that there would be any bats in the tunnels because the creatures prefer caves with high ceilings.

Like the bats, I too favor more open caves, probably because an experience in that cramped cave prompted one of the few times in my life where I experienced the physiological stress called *The Emergency Syndrome.* As I was crawling through those tight tunnels to find bats, I suddenly found myself stuck—I could move neither forward nor back. Trapped, I experienced the racing heart and tense muscles described by Cannon in 1915 as the body's "organic preparation for action," otherwise known as the *fight-or-flight response*. This is not to give the impression that while firmly wedged in an underground tunnel, the physiologist in me was calmly assessing this syndrome. It seemed an eternity, though it was probably only minutes before, to my great relief, I finally managed to squeeze myself backwards and out of that tight spot. In all the times I have crawled through caves to band bats, and there have been many, that was the only time I felt fear.

The hillside cave was a different story—it was spectacular! As I later wrote: *"The Indian Oven Cave is an extensive limestone fissure on top of a high ridge. It has two tall chambers, approximately 50 feet in height, connected by a series of small chambers. Over the floor of the cave runs a brook, disappearing as a waterfall in a low, wide chamber at the end of the cave...about 150 feet from the entrance"* (Folk 1940).

A local man who farmed on the school's property told me about this Indian Oven Cave, and when I discovered bats hibernating there, I wrote to Don Griffin. Don soon appeared with a crew, including my old friend Garrett Eddy, to band all the bats in the area, including those I hadn't reached in the Twin Lakes Cave.

Every week that winter, I visited the cave in New York and began to collect data on those bats; this exciting activity helped shape my future academic plans. I was determined to continue my education once I had saved enough money, and wanted to begin working on technical papers as soon as I could attend graduate school.

Charlotte Quaile was very supportive and loaned me her Packard for the drive through the mountains to the cave. Then came the fun part, because the easiest way to get up the large hill—and then back to the car—was to use skis. I enlisted some of my students to help in banding and observing the bats. It was the first experience they had of research in the field, and was a great way to learn biology.

During my fifteen trips to the cave, I found interesting fluctuations in the bat populations during the winter, and I soon had the material I would need to write and publish my first technical article once I had returned to graduate school.

## Rock Climbing Adventure

When I'd told my undergraduate advisor, Jack Welsh, that I was going to Salisbury to teach he had been quite excited. Jack knew the area intimately because he had taught for three years at the Berkshire School, which is just over the state line in Sheffield, MA. The distance between the two schools is only about 12 miles.

Jack had described a steep, rocky mountain behind Berkshire School where he had found dens of hibernating rattlesnakes. He was talking about Mount Everett, which at 2,624 feet is the second highest peak in Massachusetts. (The highest, 3488-ft Greylock, is much farther north.) Being a biology teacher, Jack had captured several of the rattlesnakes to use in anatomy lessons to show his students how the snakes injected their poison.

I always enjoyed hiking in the rugged area around Salisbury, where a section of the new Appalachian Trail had recently been marked, but I had never been rock climbing. That changed the weekend my old friend Garrett Eddy came to visit and explore the area where I lived. As one of the aforementioned *master bat banders*, Garrett had been to the area's caves; on this trip his passion was for climbing, and he'd asked me to show him some cliffs.

As I thought about places that might offer good climbing, I remembered Mt. Everett behind the Berkshire School. I had heard that on a clear day from a fire tower on its summit, one could see mountains in five states: the Catskills in New York, the Berkshires in Connecticut and Massachusetts, the Green Mountains of Vermont, and Mt. Monadnock in New Hampshire. We headed for Mt. Everett.

We set out on a hiking trail that passed cascading waterfalls before it rose steeply and we found ourselves leaving the lush greenery and looking upon the eastern face of Mt. Everett. As we began a climb of about a mile through the challenging, rocky terrain, I remembered something that gave me pause: this was where Jack Welsh had found all those dens of rattlesnakes!

Suddenly, I became aware of all the rocky crevices and sunny ledges that could be harboring poisonous snakes. If I had prepared myself for an encounter with a rattler, or set out with the intention of finding them, it would have been different; my scientific curiosity and careful precautions would have allayed nervousness. But that day it was different.

Try as I might to notice the changing flora—how the trees and bushes began to be stunted as we neared the summit—I kept recalling Jack's article on rattlesnakes and how they were captured in these very cliffs. Worse, I could not shake the image of Jack's photograph used to demonstrate the hollow fangs of pit-vipers: the skull of a snake, jaws stretched as if to strike, a wire protruding from the hollow fang to demonstrate the venom-delivery route.

I tried to be casual as I remarked to Garrett that there might be rattlesnakes in the area and cautioned care. But we met no rattlers during that climb. At the top of the trail, I regained my perspective, and was able to enjoy the fine view.

## How I Became an Episcopalian

Because my father was a Methodist minister, people sometimes ask how it was that I came to be an Episcopalian. "Because of beer!" I sometimes answer, relishing their startled looks. Of course it's not the full answer, but it is true in part, and here's the story.

Salisbury was an Episcopalian boys' school. In the student dorm, there were three floors with at least one faculty "monitor" on each. On my floor, I shared monitoring duties with the Reverend Bob Hulbert, who became a good friend. By the end of the day when the students were supposed to be in bed, we both grew weary of the chatter of young voices and would go into town for a beer. The town had a hotel and bookstore, and each day after the bookstore closed, it became an after-hours club where the friendly owner served highballs and beer. At the bookstore, Rev. Bob and I often discussed the history of religion, which I enjoyed very much.

I had never been a drinker in college because I needed to have my wits sharp to work, study and pass the difficult biology courses. I also hadn't imbibed at home because Methodists, at least the ones I knew, did not drink. Often enough when I was a youngster I'd heard the adults joke: "Too bad I'm a Methodist and can't drink!" But on those evenings at the bookstore-turned-club with my friend Bob, I found that tipping back a glass or two was enjoyable. "And that," I tell my friends, "is when I started my drinking career."

That year began my interest in the Episcopal Church. I went to Episcopalian services while at Salisbury School, and again later when I taught for several years at another Episcopal prep school in Massachusetts. I knew I liked the services, philosophy and approach to religion.

I wanted to discuss this with my Methodist-minister father, but I was a bit worried; it was not just that he was famous for his Welsh temper, but also that I would be leaving the brand of Protestantism to which he had devoted his life. Finally though, I worked up my nerve and approached my dad. After giving him a brief synopsis of my reasons, I asked, "Do you mind if I join the Episcopal Church?"

My father's answer was a happy surprise—and a bit of a shock.

"Son," he confided. "I always wanted to become an Episcopal Priest. You go right ahead!"

That was something I had never known about my father.

## Wassaw Island

That spring and the next year as well, the Quailes invited me to spend the three-week Easter vacation with them on Wassaw Island off the coast of Georgia. To a New Englander at the end of a long winter, luxuriating in the sweet and mild air of Wassaw Island was a visit to a tropical paradise. Strolling down the main avenue to the sea, one's view is framed by trees hung with blowing Spanish moss. The expansive white beach is margined by palmettos on one side and rolling white surf on the other, with a background of deep blue sky.

The small, seven-mile island belonged to Charlotte Quaile's uncle, Henry Parsons, who was also editor of the *New Republic*. Besides Mr. and Mrs. Parsons, several others were also there to enjoy the warm sea breezes: the Quailes' son George, Charlotte's brother George Milmine, and another stimulating guest, Phillip McMaster, a professor from the Rockefeller Institute.

**Wrangling an alligator on Wassaw Island. *L to R:* Milmine, Folk, McMaster.**

The island was full of wildlife, from seabirds to deer to alligators, and my visits there crystallized my desire to study natural history. I had my first experience with poisonous water moccasins, collected some charming, three-to-four foot long kingsnakes that were quite tame, and I made study skins of the birds felled by Emerson Quaile and another hunter. Apparently, no ornithologist had visited the island, so, as *de facto* naturalist, I made notes of the many bird species I saw. My experiences there over two spring vacations provided me with the material for my first published paper, "Easter on Wassaw Island," which appeared in a regional ornithological journal (Folk 1939).

Hoping to catch an alligator, we went hunting for them in the evenings. In the end, Milmine, McMaster, and I succeeded in getting a noose around the head of a 6-foot 'gator, which we put in the swimming pool so we could photograph it the next morning. After we got our shots, we let the fellow go.

But the most unusual experience on Wassaw Island had to do with a young bald eagle. I'd always had a passionate desire to photograph baby eagles in their nest. At either end of the island stood two large Southern Jack Pines with nests that still had young eagles. These large evergreens had trunks like walls and no branches low enough for a person to grab, and I knew it would be nearly impossible to climb the tree. Even with the help of a man who worked for the owners of the island, I not only was unsuccessful in reaching the lowest branches, but also managed to nearly kill myself by falling—I probably would have if not for the providence of having that man along with his good reflexes and strong arms to catch me!

I still wanted some pictures of baby eagles, so after thinking the situation through and gathering some tools and supplies, I went to the nest at the other end of the island in hopes of a better result. I nailed pieces of wood to the trunk of the massive tree to make a ladder, which I finally managed to ascend. Eagle's nests are huge, spanning six feet and weighing several thousand pounds. Eventually, I made it to the nest, and there I was perched on a branch beside it and delightedly taking pictures of the babies, when to my horror, one of the young eagles hopped out of the nest and fell to the ground; fortunately its fall was broken by the flapping of its wings in a frantic attempt to fly.

**Young bald eagle at the cottage on Wassaw Island.**

The uninjured bird let me carry it back to the cottage, where everyone was of course entranced by the strikingly beautiful young raptor. I hand fed the eagle as it perched on the porch rail, and it seemed content to just hang around the cottage.

However, since everybody was to leave the island within a few days, we were concerned about whether the immature bald eagle could survive on its own, but we knew we couldn't possibly take our national bird on the train with us back to Salisbury! But here is where the resourceful Charlotte Quaile used her imagination and spunk to solve the dilemma.

"Of course we can take it back with us on the train," she said, "we'll simply call it a chicken!" Thus, along with our baggage, one "chicken" was shipped on the train from Georgia to Connecticut.

For a short time we kept the eagle in one of the large barns on the property. I fed it fish by hand and fortunately knew not to overfeed it, which is a temptation for people rescuing wild creatures, but which can make the animal sick. Our eagle soon was flying much better, sometimes banging into the barn's walls and roof. It was now wearing a government bird band.

At that point the "eagle committee" decided it was time to release it. The young bald eagle had been quite a celebrity during its short tenure at the school, and the entire faculty and student body gathered on the front lawn to watch its return to the wild.

**Folk with the "chicken."**

I'll never know if letting it go like that was the best thing to do for the eagle, but I hope it was. My last sight of that bird is a picture I'll never forget. As I held my arm aloft, the beautiful raptor extended its lovely wings, rose into the air, and soared over the lake. Everyone watched until it disappeared into the evergreens on the far shore.

At the end of the school year, I had saved enough from my paychecks that I could apply to graduate school, although I regretted having to leave this place where I had enjoyed such wonderful times. Because my life there was so idyllic, I feared that if I didn't return to school right then, I might never do it.

Although I'd always planned to continue my education at Harvard, I had also received an offer from a graduate degree program at the New York Museum of Natural History. A fascinating researcher there had published an article reporting that the pits in the heads of pit vipers are temperature-sensitive sense organs. I had traveled to the museum to visit him, and because of my background in herpetology, he had invited me to be his graduate student.

I chose instead to return to Harvard, where I could continue under the sponsorship of my earlier mentor, John Welsh. I had a very sad parting with the Quailes, although I knew we would remain good friends, even at a distance.

**The eagle takes flight from Salisbury.**

I approached graduate school as if it were a sumptuous banquet table where it is hard to decide which of the attractive foods you want to select: in what area of Biology did I want to specialize for the rest of my life?

# *Chapter Four*

## *Graduate School*
**1938–1940**

I RETURNED TO Cambridge in 1938 and found myself in a group of graduate students who, from my vantage point, seemed to be the most stimulating people one could see on a day-by-day basis. While some of us had banded bats with Donald Griffin as undergraduates, I met three new students who later would become leaders in their fields: Peter Morrison, George Bartholomew, and Oliver "Paney" Pearson. All of us preferred to be collecting information on animals in the outdoor environment. We could never get enough of a blue sky and billowy clouds, birdsong, or the intoxicating odor of dense wet woods. In my own case perhaps the strongest emotion was engendered by the snow on hemlock branches near the top of a mountain.

But as my academic schedule began, I found I had decisions to make. Along with difficult courses such as Physiology, Genetics, and Botany, it was necessary to decide which area in biology would be my specialty. I was still influenced from my undergraduate years by the fascinating ornithologist Elliott Ludlow Griscom. He had led a number of us to do field identification of birds, and fostered a desire to understand what influenced bird migration. I also had made some bird museum specimens for him, always from birds killed on the highway by cars.

Meanwhile, even while I was taking courses in the biology department I maintained a friendship with several people in Harvard's Museum of Comparative Zoology (MCZ). One was my undergraduate professor Phil Darlington, the taxonomist and biogeographer who had inspired me to trek across the Gaspé Peninsula to collect insects for his research. Another most enjoyable person at the MCZ was Glover Allen, from whom I always received enthusiastic encouragement. I gave him study skins for the museum that I'd made of mammals from Wassaw Island, where I'd spent time with the Quailes.

Those interactions and experiences made me wonder whether I should be a biogeographer. But I had one undercurrent of thought, which was that I was actually becoming more interested in investigating the *physiology* of birds or mammals or reptiles, and not just in identifying them. I had an interest in studying their normal mechanical, physical, and biochemical processes, whether on the level of the whole organism, or on the level of its cells, tissues, organs and other systems. And my curiosity always came back to wanting to know more about the interaction between the organism and its physical environment.

Interestingly, I found that Pearson, Morrison, and I all were discontent with just doing course work and taking exams and wanted to be carrying on experiments at the same time.

## Imaginative Student Investigators

One of the most motivated of these students was Peter Morrison, who, when he wasn't studying, trapped small mammals on Mt. Washington and brought them into the laboratory to measure their metabolism. Eventually, Peter studied mammals and insects in the same fashion in Australia and in many parts of South America. Much later, and much farther north, he became Director of the Institute of Arctic Biology at the University of Alaska, Fairbanks. While there, he spearheaded construction of a new building dedicated to one of his early mentors, the great scientist Larry Irving.

Another student, Oliver "Paney" Pearson, had the same habits as Peter, running a small mammal trap line and bringing the results into the laboratory. Paney had stumbled upon the most exciting experiment of all after he had read that the saliva of shrews might be poisonous. Making an extract from the salivary glands of the Northern Short-tailed shrew (*Blarina brevicauda*), he had been tempted to inject it under his skin to see if it caused inflammation. Fortunately, he did not, but injected it into mice—which died immediately. As he studied this poison the next day, he found it to be drop-by-drop as poisonous as cobra venom. As another surprise, he found that *Blarina* was the only species of the many he had collected that had this poisonous saliva. Over time, this may have been one of Paney's most important discoveries, and this work was quickly published in a refereed journal. Not long afterwards, he measured the metabolism of hummingbirds flying in a bell jar. I believe that these measurements showed one of the highest oxygen demands known for any vertebrate tissue.

After some time, Paney Pearson left graduate school to go down the Amazon River and collect specimens for various museums in the United States. When I asked who his companion would be his answer was, "He's a hundred-yard Harvard man," the somewhat obscure meaning being that one could detect from a distance that he was from Harvard. Actually, his companion was the son of the individual financing the expedition.

One of the most exciting reasons for Paney's expedition had to do with an important discovery by Professor F. L. Hisaw. Hisaw had found that before birth the guinea pig releases a hormone, which he named *relaxin*, which causes the two parts of the female pelvis to separate and become moveable at the time of birth of the young. Paney Pearson was to capture and study some ancestors of the guinea pig in the highlands of Peru and to determine whether the parts of the pelvis were separated at parturition. It is my understanding that the event which the biologists wished to witness in this ancestor of the guinea pig did not happen. To follow out the career of Oliver Pearson, he eventually became a professor at UC Berkley and then director of their Museum of Vertebrate Zoology, maintaining frequent research trips to South America.

George Bartholomew was working with Prof. Clark, the only animal ecologist in Harvard's Department of Biology. Clark had been a very inspiring teacher to all of us, and we gathered in his seminar to exchange ideas under his guidance. Bartholomew, or Bart, as we called him, was studying photoperiodic influences upon English Sparrows and had a fascinating thesis, which consisted of finding what parts of the color spectrum were responsible for the enlargement of gonads in birds. Unfortunately, one night while collecting experimental subjects from vines on the outside walls of the biological laboratories, he fell off his ladder and sustained injuries that delayed his graduate work.

Bart went on to become a professor at UCLA and head of the Department of Biology. I was very inspired by some of his future field work, which became classics in biology. In one of them he described the metabolic profile of the responses of fruit bats to different ambient temperatures. Even more exciting was his study of the kidneys of the aquatic iguana, which he pursued on remote Galapagos Islands. Most iguanas live in deserts or tropical heat, not in water. The evolutionary process caused a different form of kidney to develop in the marine iguana because of its saltwater home.

Amongst this group of graduate students was a prolific researcher who was developing a thesis on the influence of environment on physiology. In his excursions into the outdoor environment, Charles Lyman had become fascinated with the color change of snowshoe rabbits in winter. He selected as a thesis the influence of light, a photoperiodic effect, on the seasonal coat change of the snowshoe rabbit. After graduating, Charles was one of the first to realize that the golden hamster, an animal common in pet shops, was a good hibernator. Later, from the Department of Anatomy at Harvard, Charles published several hundred papers about mammalian hibernation and organized the first international conference on the subject.

Another person who depended upon the outside environment to provide his research material was Ralph Smith. He traveled along the Charles River to empty his crayfish traps each day to provide material for his thesis subject. He discovered a hormone that controls growth and shedding of the exoskeleton of the crayfish which became known as the "eyestalk" hormone. Years later Ralph continued his study of aquatic invertebrates in the outdoor environment along the Pacific shore. While he was head of the Department of Biology at Berkley, he rewrote and added to the premier field manual on that subject, *Light's Manual: Intertidal Invertebrates of the Central California Coast.*

Lunchtime was a period of stimulation and exchange of new ideas. In the combined office/laboratory I shared with Ralph we had a device on which we could make toast. The procedure was to toast one side of two pieces of bread and then to pile on jelly and peanut butter. We never varied this menu because its consistency saved time.

Frequently we were joined by Don Griffin, who while still a graduate student had received a research grant to study bird migration. I recall vividly the excitement in his voice one day when he announced, "Today I learned that Von Frisch in Germany has described and understands the language of bees. I consider this as important as the discovery of the origin of species." It was very unusual for Don to make such an extravagant statement; always a brilliant force, his was a silent power. And when he spoke, people listened.

In his bird migration studies, Don was interested in determining which senses birds use when they migrate, often over long distances. There had been many experiments in which birds were taken a distance from their nests and then released, but Don suspected there was a flaw in their design—because the birds had been transported in a linear fashion, they might have remembered a map of the direction in which they were taken.

Don, of course, had a plan to resolve that problem, and I was to learn about it when Don burst into my lab one rainy night while we were both working late.

"I need your help," he said. "Can you come with me now?"

I knew it was important from the way he asked, so I left my bench and headed outside with him while he explained the situation. He led me to his car, attached to which he had an odd contraption: it was a platform on wheels that had four wooden boxes that were slowly revolving. Each box contained a female herring gull taken from its nest on Penikese Island in Buzzards Bay near Woods Hole, MA. The platform and boxes were revolving because Don wanted to make sure the birds didn't memorize the route.

I was to drop my work, drive with him to a mountain in Montreal and help him study the birds as they were released. Of course I said yes.

The funniest part of that long trip was when we had to stop for gas on the way to the mountains. It was about 3 am, and I'll never forget the expression on the attendant's face when he came out to put gas in the car and saw the rotating boxes on the trailer—he wore a look of horror as if he thought we were going to blow up the station.

It was daylight when we got to the mountain. Don's plan was that he would hike to the top of it and I would remain at the bottom with the birds. Then, upon getting a signal from him, I was to let the birds go.

Things would have gone smoothly, but when Don was in position and gave the signal, I realized I didn't know how to open the boxes. I hadn't seen how he made them, and I was trying to take the tops off because it looked like the logical way. Instead, however, Don had nailed them loosely at the bottom to come apart at the tray. Up on the mountain Don was watching through his telescope, thinking, "Ed is ruining my boxes!"

The day was saved when one of the boxes slipped a bit and I could see that it opened from the bottom. Amid no fanfare, I let the gulls go, one by one.

Don was observing the gulls from atop the mountain with his powerful telescope to see which way they headed. Would they need to circle around to get oriented? Would they head east and then follow the shoreline? Or would they fly straight back to their nests? The answer was that they headed in a straight line to the shore and then followed the coast down to their home island off the coast of Massachusetts. They were back on their nests before Don could get there to log them in.

Another graduate student in our group spent so much of his time outdoors with an insect net that it was no surprise that Carl T. Parsons became a dedicated entomologist. In recognition of his work on a difficult genus of beetle, *Coleoptera Nitidulidae*, a new species was named after him: *Epuraea parsonsi* (Connell 1981). He spent part of his later career curating at the Smithsonian.

Donald Fawcett was my bench mate when we took our first histology course. Don was a pioneer in the use of electron microscopy and he wrote the most widely-used textbook of histology (Fawcett 1997). After receiving his MD degree at Harvard, he served as a battalion surgeon in WWII. He then returned to research, and eventually became head of Anatomy at Cornell University as well as at Harvard. His interest in veterinary medicine led him to Africa, where he studied parasitology as director of electron microscopy at the International Laboratory for Research in Animal Diseases in Nairobi, Kenya.

There was another group of graduate students who restricted their technical activity to the indoor environment. Carol Williams made his name by discovering the hormones controlling the life cycle of large moths. Another, Lee Chadwick, was an excellent insect anatomist.

Vincent Dethier was an insect physiologist who taught me how to do delicate operations on insect larvae. He became a respected entomologist who wrote books and articles in the sciences and in many other genres as well. I always remembered him because later on he took time out from career to write and publish a children's book about their family's pet duck.

One of the most interesting "indoor biologists" was Marcus Singer. I remember my conversations with him on New Year's Eve of 1938. I was to join my family the next day, but I was working on a paper for publication and wanted to get in a long quiet evening of composing. As I approached the biological laboratories, a vast building in a large L shape, I was surprised to see only one other light in the entire building. It was near my office and turned out to be the laboratory of Marcus Singer. We had talked occasionally about his work with Leigh Hoadley, the embryologist whose crockery pots full of rotten eggs I had emptied in my freshman year.

At any rate, Marcus and I decided to go out at 11:30pm to celebrate the New Year. We drank some beer and he told me about his exciting experiments which were simply this: as an embryonic nerve is becoming established in a developing limb it moves forward like the arm of an octopus. Marcus had found that at the forward surface of the active growing embryonic nerve, a hormone was produced.

Later—for reasons that were decidedly unfair—Marcus had to stand before Congress in Washington because, simply out of curiosity, he had attended a communist meeting. Thanks to his successful career as an embryologist, his students and fellow researchers formed a Singer Society to honor his name.

Another person wedded to his laboratory was Nevin Scrimshaw. He also worked with Leigh Hoadley to describe in detail the development of differences between fish that laid eggs and fish that gave birth to live babies. More important was his later work; he became an MD and food scientist who worked tirelessly to improve nutrition in malnourished populations, especially in South America.

The person we went to for advice was Hermann Rahn, the post-doc I had taken to a den of garter snakes while I was an undergraduate.

As a physiologist, Hermann had a broad range of interests, including the physiology of responses to high altitude. Most physiologists today are familiar with the Rahn-Otis High Altitude diagram used to assess acclimatization, as well as his pioneering work on the physiology of the women pearl divers, the Ama in Korea and Japan. Later he became known as an outstanding field ornithologist as he studied the respiratory physiology of avian eggs. But Hermann's initial interest had been the physiology of reptiles.

During those two years I was in graduate school, I enjoyed a most inspiring relationship with my sponsor, Jack Welsh. Whenever the graduate students in Jack's laboratory were having lunch together, he was apt to join us. There were two things that emanated from his conversation. One was the importance of nerves as secretory glands; the other was their products, neurohumors. Later the more common word was "neurotransmitter," and Jack proved himself to be an outstanding futurist in their study.

**John H. Welsh. Outstanding scientist and wonderful mentor. (c. 1992. Photo courtesy of Welsh Family.)**

Before becoming fascinated with neurotransmitters, Jack Welsh had specialized in studying biological clocks, or "circadian rhythms." As beneficiary of Jack's knowledge of biological clocks, I was studying how hours of daylight triggered the season of reproduction. For example, in the North Temperate Zone there are large changes in the light cycle so that in the city of Fairbanks, Alaska, there are only two or three hours of daylight during winter, compared with areas near the Equator, where every day for the entire year there are twelve hours of light and twelve hours of darkness. Part of my work involved trying to understand how hours of daylight change the time of breeding in various mammals in the free environment, particularly bats.

## My First Technical Paper

I was writing the story of the restless hibernation of New England bats in their winter caves. When I read articles about the influence of hibernation on reproduction, I realized that there was a mystery described in the literature. A scientist named Mary J. Guthrie had published several papers on the endocrinology of the reproductive tracts of bats; several times she mentioned a curious phenomenon about their reproduction. The males were ready for breeding in the fall, but the females do not ovulate until they leave hibernation in the spring, which is when the pups were born. No one had learned, in a reproductive sense, what happened in the interim of three or five months before the young were born. Was there mating during hibernation? It had been established in the literature that in some regions, such as Missouri, bats do copulate throughout most of the winter. Professor Guthrie had studied over four species of bats, and never found fertile ovaries in the fall. This led her to suggest that the only logical answers to this reproductive question must be either that in February, spring inseminations have started, or that the female stored spermatozoa all winter long, which she did not believe was the case.

I determined to settle the question. Although I was taking three difficult courses (mammalian physiology, genetics, and embryology), I decided to do the critical experiment. It had a very simple design. I would capture hibernating females, maintain them in hibernation all winter, and then study the histology of their reproductive tracts in the spring.

Upon Don Griffin's advice, I used *Eptesicus f. fuscus*, the big brown bat, because of its hardiness for such studies. That winter, I traveled to Ely Mine in Vershire, VT, where I gathered five hibernating female bats. For the trip back to Cambridge, I packed them carefully in a container surrounded by ice, and then isolated them in a cold room in the Biological Laboratories at Harvard. Isolated from all males, they hibernated until April, when I brought them to normal temperature and fed them by hand.

Hand-feeding the bats was standard procedure after successful hibernation; other scientists had discovered that this was necessary to maintain these small creatures in captivity, since ordinarily they would catch their food—insects—in the air. This by itself was an exciting experience, and it was amazing how remarkably tame the female bats became. I did not have to hold them in order to feed them; they simply learned to come to a dish and feed themselves. We gave them two kinds of food: live mealworms and honey bees in cream cheese. Soon we were able to release flies in their cage for them to catch on their cage wall.

A few weeks of feeding the isolated female bats allowed them time for the reproductive cycle to be completed. I euthanized the first one and examined the reproductive tract. You can imagine my excitement when, there in the uterus, was a partially developed embryo. The other females showed the same result. I excitedly wrote up these results and took them to Leigh Hoadley, the expert on embryology. I had of course made slides of the fetuses, and photographed the slides. He was just as excited as I was, and there was much investigation as to whether there could have been a mistake. Fortunately, John Welsh, Hermann Rahn and Donald Griffin had followed all the details of my experiment, and my paper, "The Longevity of Spermatozoa in the Cave Dwelling Bat," was quickly accepted. It was my first technical publication.

I had proven that the spermatozoa had lasted in the female bat for three months as viable sperm cells, since they were able to fertilize the ova produced in the springtime. Assuming that the females did not ovulate until taken out of hibernation in April, the survival of sperm in the tract was considered to be ninety days. This phenomenon in non-mammalian vertebrates is not remarkable. It was well known that turtles can store spermatozoa for as long as a year, and even female chickens can retain viable sperm for a month or more. However, this was the first case that showed that sperm could be stored in a female *mammal*.

Although I felt very pleased at the results and their rapid publication, I was quite surprised at the interest among embryologists around the world. I received complimentary letters from as far away as Oxford, England. I think that the medical profession had not yet begun to freeze human sperm, and I presume some of the people who wrote congratulatory letters were interested in that field of study. It was most inspiring to a young graduate student to receive these letters; during my later years of teaching I attempted to remember their effect and write similar letters to young graduate students.

## Botany

We were all required to take a course in botany. I satisfied half of the requirement by attending the exciting lectures by Prof. Cap Weston. This led me to later teach about the fascinating world of *motile gametes*, or reproductive cells, of plants. Some plants have spermatozoa and have evolved sexual life cycles just as animals have, although with some plants, the motility may come from being dispersed over the countryside by the wind.

For the second semester requirement in botany a group of us, including Don Griffin, signed up for a course in forest ecology given by Hugh Raup. He was an inspiring professor who conveyed the impression of a teacher sitting at one end of a woodland log smoking his pipe while his students listened to him as they too sat upon the log. While taking the course I met a young botanist on a fellowship from England, Eric Watson, who became a lifetime friend.

Prof. Raup taught parts of this course by taking the class in cars to distant places to demonstrate the kind of natural forests that existed throughout New England and the eastern United States. In most cases we spent several nights in motels on the road. Raup had done his field work in the High Arctic and Greenland, and during evenings spent in motels he stimulated our interest in visiting those places with interesting and useful anecdotes from his time there. I believe his research methods of study were at the time unique in forest ecology.

For example, he would search out town or county records to look at the deeds, which were dated in the 1700s or 1800s, in order to find descriptions of the removal of local forest cover.

## A Summer Job and Important Friendships

For several summers, I earned my graduate school expenses by working for the Prescott family of New Bedford, MA. There were several adult sons and daughters whom I got to know very well over three years. I was initially hired as a companion and tutor to the grandson, Conrad, who was being raised by his grandmother because his mother was absent, living somewhere on a yacht most of the time.

The first summer Conrad needed to study hard to be able to pass an exam that would allow him to move from $6^{th}$ to $7^{th}$ grade. I made up a class schedule for each day and slid it under his door—I even illustrated it because I loved to draw and felt it might lighten the blow of his having to study all summer. Unfortunately, Conrad didn't like to study at all, and the subject he hated most was arithmetic.

One day, he came at me with an ax, yelling, "If you make me do that arithmetic, I'll hit you with this ax!"

"Then I'll have to tell your grandmother," I threatened back while stepping carefully out of reach of the ax. That scared him because his grandmother could be very fierce and was always after him for things like not having his shoes polished.

The second summer, I taught Conrad to swim, which was much safer than teaching him arithmetic.

The Prescotts had a large dairy farm called "Maryano," with a grand manor house that had white columns in the style of a southern plantation. During the three summers I stayed there, the Prescotts made me feel like a member of the family. One son, Bryant (Bry), became a lifelong friend and sometime travel companion. His brother, Oliver, was a lawyer in New Bedford and lived in another house on the property. Bry, Oliver and I shared a love of hiking and had long conversations about the trails in the White Mountains.

I became good friends with the Prescott's daughter, Lydia. She was married to a wonderful chap named Gordon, and they had an apartment in Boston. But Lydia was staying at the family farm because she was recovering from tuberculosis. Before I knew her, she had been in a hospital in Arizona, where, as fantastic as it sounds, the treatment was to collapse the lung by inserting sterile Ping-Pong balls in the pleural cavity between the lung and ribs. This is known as *plombage therapy*, and was based on the theory that collapsing the lung would allow it to rest and thus heal the lesions caused by TB. (Later development of antibiotics made this treatment obsolete.)

**On one of many horses I rode; sometimes it was part of the job.**

While Lydia was recuperating at Maryano, her mother thought riding would be good for her and "help build her back up." I was given the delightful assignment of buying her a suitable horse. I traveled to Worcester, about 65 miles away, where I knew of a good stable. There I selected a lovely palomino gelding and had him shipped by train back to New Bedford. Lydia adored that horse.

Lydia's parents didn't want her riding alone, so they rented a horse for me to ride that summer. I loved being on a horse again so much that I rode out bareback on that lively western horse by myself every morning before breakfast.

Almost every weekend, Gordon came to New Bedford to visit Lydia, and when he did, we all went out on the town. There was an empty guest cottage on the property, and sometimes in the afternoons we young adults would gather there to play music and dance. That's how I met a friend of Lydia's, Dorothy Dodge. She was beautiful, fun, and unattached, and had a young son named Christopher. By the time I had finished graduate school, Dorothy and I were spending a lot of time together.

I received my MA degree in 1940. My preference was to apply right away to the PhD program at Harvard, and the faculty in the

Biological Laboratories encouraged me to do just that. However, *The Financial Effect* dictated otherwise. I needed some form of assistance and unfortunately, in those days there were scarcely any scholarships for graduate work. To earn a living as a graduate student, one could become an instructor in laboratory courses, but since these positions were very popular, I could not get one. Therefore, I knew I would need to find a teaching job elsewhere to earn money enough to return to graduate school.

# Chapter Five

## Science in the West Indies
### Summer, 1940

### Field Work on the Far Horizon

JUST THEN, AS I was searching for a job, a very pleasant thing happened: John Welsh arranged for me to do field work to fill in the time before I was to leave to start teaching. Jack had spent some time doing research at a biological station on the southern shore of Cuba, about 180 miles from Havana. He had enjoyed his own research there, and published one or two papers from his studies. Because he knew I would find it exciting and profitable for my own work, he secured a fellowship for me at the Soledad Laboratory of the Atkins Institute in Cienfuegos, Cuba.

This tropical research station had been set up on a plantation owned by American sugar planter Edwin Atkins to conduct sugar cane and crop research, and grew to encompass a tropical garden with thousands of species from around the world. It became part of Harvard University under a large endowment by Atkins, and Harvard administered it through its Arnold Arboretum in Boston; thus it became the Atkins Institute of the Arnold Arboretum. (After the Cuban revolution in 1959, the arboretum became the Cienfuegos Botanical Garden, a Cuban National Monument open to the public.)

In the 1930s and 40s, this remarkable biological station sponsored botanical and zoological research of professors, graduate students working on fellowships, and visiting fellows. Harvard's director of the Agassi Museum in Cambridge, Thomas Barbour, had helped develop and promulgate the biological station and the 22-acre tropical garden.

The problem I chose for my research in Cuba evolved naturally from my earlier work on the reproduction of bats. In that study I had shown that spermatozoa remained viable in the hibernating female for three months and could fertilize ova produced in the spring.

The biological question still remained, given the history of migratory bats and the seasonal changes brought about by the ice sheets. Was it paleontological events that caused this curious storage of spermatozoa in the female bat? In earlier times, was the regular breeding time of migratory bats in the fall or in the spring?

Thus, the question I presented when I applied for a scholarship to the Atkins Institute was the following: *What time of year is the breeding season of bat species near the equator?* Do they have a specific breeding time in spring or fall, or do they instead breed at random throughout the year because there's no cold weather to influence their behavior? In addition, at the equator, the hours of daylight for the entire year are twelve hours of light and twelve hours of darkness. Thus, what role does daylight play?

Essentially a study of biological clocks, my project involved trying to understand how hours of daylight change the time of breeding in various mammals in the free environment. Therefore, my mission would be to study the caves of Cuba to find out during the two-month visit which bats were breeding, and then before I left the island, to make arrangements for the caves to be checked each month of the year during my absence.

Then another influence modified the plans for this expedition to Cuba. I still spent hours of conversation whenever possible with Phil Darlington, the young professor and biogeographer for whom I had once collected insects on the Gaspé Peninsula. When Phil heard that I was going to the Caribbean, he arranged for an additional small fellowship for me to go to Haiti. In order to develop his ideas for his forthcoming book, he needed more information from Hispaniola, the Caribbean island containing both Haiti and the Dominican Republic. However, I wasn't sure I could go because Phil's fellowship would pay only part of the cost, and as a very poor graduate student, I could not finance the rest of the trip to Haiti.

Much to my delight, Emerson and Charlotte Quaile from Salisbury were so interested in these stories of field biology that they financed the side trip to Haiti. Because there had been many revolutions there, scientists had not been able to collect museum specimens to the extent that they had on other islands. Phil made arrangements for me to go for two weeks from Cuba to Haiti to collect material for him. The target location was a plateau in the *Massif de la Salle,* which spans much of Haiti's southern peninsula and runs east into the Dominican Republic. In this range is Haiti's highest mountain, Pic la Selle, which at 8,793 ft is higher than "my" Mt. Washington (6,288 ft), the tallest mountain in the Northeastern United States. Phil wanted me to go into the interior, climb one end of this plateau, walk its length, and come down the other end, all the while collecting insects and reptiles for him. These specimens were to be markers to help Phil understand the evolution of vertebrates and their distribution over the Islands of the West Indies.

Phil helped make me into a field biologist with advice like this: *"Remember Ed, you don't need to try to be too comfortable when you go on these sorts of expeditions. For example, if you have a camp and you see another mountain peak nearby, just take a blanket and climb that other mountain peak and collect the animals there and then spend the night. Sure, you might shiver all night, but it doesn't hurt you any, and then you can come back to your own camp and recover."* I felt nothing but excitement at the thought of such an experience.

Yet another major influence came into this expedition when Hermann Rahn heard I was going on this trip to the West Indies. Knowing that I would be collecting *Anolis* lizards for the Smithsonian Institution, Hermann told me what he wanted.

"Well," he said, "as you know, I've been finding a mammalian type of endocrine material in reptiles. Some reptiles' ovaries appear to have the *corpora lutea*, which make a special hormone called *lutin*. You can bring me some reptile ovaries to study."

In all the reptiles I collected in Cuba and Haiti, I was to open the body after killing the reptile, and if it were a female, put the ovaries in a fixative solution so that slides could be made of the organ. By this means, his survey of looking for *corpora lutea* in reptiles could be extended. I was only too glad to take on this extra responsibility.

## Cuban Quest

My English friend Eric Watson, from Dr. Raup's forest ecology course, had decided to accompany me on part of the trip to Cuba. Eric was an accomplished ornithologist, and together we had visited many exciting birding areas in New England, such as Great Blue Heron rookeries. As we made our way through the southern US towards Cuba, we visited the famous Okefenokee Swamp in Georgia, and later took a boat trip down the Florida Keys with a member of the Fish and Wildlife Service. In both places we were able to photograph some unusual birds in their nests. We also stopped in Miami at the Fairchild Gardens, an exciting arboretum with the world's best collection of the oldest seed-bearing plants on earth, the *cycads*. The director asked us to take some special seeds from India to the Atkins Institute in Cuba.

Then Eric and I boarded a ship to Havana. I wanted to arrive in Cuba ready to go to work; I knew that there would be considerable need to preserve specimens for the various museums I was responsible to. Since 70% ethyl alcohol was very cheap in the laboratory, I had brought a gallon of that along with my other equipment. When we debarked in Havana, it turned out that the Cuban customs officers were very strict. When they heard about my gallon of alcohol, they said, "Well, we must tax you $200.00." Of course I had no money for that, and I suggested they might pour it down the gutter. Off they went with it, but I suspect they probably made it into holiday drinks.

Soon Eric and I were on a bus to travel half the length of the island to get to Soledad. There we were met by a driver in an automobile and taken to the Atkins Institute. After I had arrived at the laboratory, I found that the confiscation of my ethyl alcohol by the Cuban authorities would pose no problems; to preserve my specimens, I should simply buy very inexpensive rum, which was 50% alcohol. (Months later, when I returned to Cambridge, as I unpacked my specimens preserved in rum, I soon noticed a group of graduate students gathered 'round, drawn by the pleasant odor.)

The Atkins Institute had been visited by numerous students and faculty from the Biological Laboratories in Cambridge. Two very helpful staff members ran the Institute, Superintendent David Sturrock and his assistant Frank Walsingham. Both of them had been trained at Kew Gardens outside of London, England. The Institute was situated at the edge of the Arnold Arboretum, and even those of us who were classified as zoologists spent many hours walking through the acres of exotic trees and plants.

As a botanist, my friend Eric particularly enjoyed wandering through this collection of florae from around the globe. One of the most spectacular is the Giant Sago Palm. Living under the palms were tropical water birds such as the Jacana. The Jacana, a relative of the Coot, is a strikingly beautiful bird with extremely long toes that permit it to walk over floating vegetation.

After a few days Eric left the Institute and returned to England to spend the rest of his career at the University of Reading, where I later visited him several times. Eric was not only an expert botanist and author of the influential book, "British Mosses and Liverworts" (1955), but he was also an accomplished watercolorist.

Now it was time to turn to the main purpose for my journey to Cuba: to study the reproduction of bats near the equator. It would take one year of studies in order to determine whether the bats were restricted in their breeding to a particular time of the year, or if breeding was distributed throughout the year. Fortunately, after my initial investigations during this trip, my experiment would be continued by the Atkins Institute after I left.

There was a group of limestone caves in a circle around Soledad where I could look for bats. I traveled by horseback to these caves with a young Cuban man who was assigned to me as a guide. We found that a number of them contained no bats during the two months I was at the laboratory, while other caves were well endowed with one species in particular, *Artibeus*. This was a large fruit-eating bat with a "face looking like a bulldog" and very powerful teeth; the wingspan was about fourteen inches. I soon found that these bats were breeding in several of the caves and prepared a system for study of the reproductive organs.

I needed a study of the feeding habits of this species and notes on the care of their young, so I decided to sleep overnight in a bat cave and record my observations. I packed my equipment on a horse and rode to the cave in the afternoon. When I arrived, the bats were noisily quarreling on the roof inside the cave, but not flying outside where it was still daylight. I tied the horse with a long rope to let him graze, and found a corner of the cave to put down my sleeping bag.

I had several questions, such as, were the males spending the time with nursing females or was this just a nursery and child-rearing location? How many hours were the bats away from the cave to accomplish their feeding? These questions were easy to answer by flashlight. I'm sure I occasionally dropped off to sleep on my sleeping bag, but spent most of the time making observations. My notes showed that mothers carrying their young flew out on forays to feed on fruit. At certain times, they remained attached to the ceiling while nursing their young.

I remember that night as being a busy one, not only because I was making observations, but also because of wildlife visitors. During the night I had three guests. One was a nocturnal mammal, a hutia that was about the size of a New England woodchuck. There were three kinds of hutia in Cuba and my visitor was the most common one. Another caller was a brightly colored snake looking somewhat like the corn snake of the United States and about three or four feet long. Because I knew that Thomas Barbour had studied the reptiles of Cuba for many years, I decided not to collect this particular specimen since there was little chance that it would be uncommon. The snake had crawled down the side of the limestone cave and I presume was hoping to make a meal of a baby bat.

My other visitors deserve a particular mention. In the daytime and at night I had walked over many parts of the cave and found that the surface on the floor of the cave was spongy underfoot. I had noticed that near my sleeping bag there were two footprints where my shoe had gone beneath the surface about two yards apart. Returning to that spot with a flashlight at midnight, I saw a strange sight beside my sleeping bag. At first it looked like a stream of vegetation about two inches wide, originating from one of the footprints extending for a yard or so before disappearing into the other footprint.

What I was witnessing was a procession of leaf-cutter ants, each carrying a portion of a plant that had been cut by their powerful jaws. These ants have been studied extensively; they can carry a leaf 20 times their weight. They do not eat the leaves, but take the plant matter underground where it breaks down to nourish a special fungus, which is what the ants actually eat.

The question of course arose, "where did the plant material come from?" I was much too occupied with the biology of bats to try to investigate, but the leaves obviously had to have come from somewhere outside the cave. That meant there was a long tunnel just under the surface, and I had obviously interrupted their procession by unknowingly stepping on the tunnel. Instead of excavating through my footprints, the ants apparently found it simpler to come out on the surface and then disappear back into their tunnel at the first opportunity.

It was not long after this overnight study that the director of the laboratory made a pronouncement which turned out to be eventful; he said that I should travel to the highland ranch owned by the sugar plantation. I was to go there and stay three days and study the bats of that region and bring back specimens. The ranch was maintained for the purpose of breeding oxen to pull the sugar carts. I had seen these animals at work; they were huge and appeared to weigh as much as elephants.

I had previously been given plodding, well-broke horses to get to the bat caves. However, I had become friendly with Señor Cataya, the foreman at the Atkin's Institute's home ranch, and I think because of our conversations about good horses and their gaits, he decided to provide me with an unusual mount. My guide, Felipe, appeared with two horses, the one I had previously ridden to the cave, and a beautiful large white mare that would not stand still.

The guide told me the name of this mare was "Poison" because she had killed a man. I think he said this because he wanted to ride her, although I have no doubt it could have been true. The horse was outfitted in very light reins and a shank bit, probably with a spaded-port—the type of bit usually reserved for horses who are trained to such a high level that only a light touch on the reins is required—or safe. The horse was so restless I wondered for a moment whether I should ride her, but decided to try.

Fortunately, I had what riding people called light hands, and I scarcely used the reins on this splendid horse; slight pressure on the bit resulted in instant obedience and change of direction. As for her name, if "Poison" was indeed her real name, I presume that some hard-fisted rider had mounted the horse and jerked on that tender mouth with a bit intended only for a subtle movements, causing her to rear and fall over backwards, a wreck that can indeed kill a rider.

Thus began a wonderful experience because the horse turned out to be at least five-gaited; I think she was a Paso Fino. These "Caballos de Marcha," have a proud carriage, and a distinctive smooth gait that can carry a person around the plantation or down the trail all day at varying speeds without the jarring experienced by the rider of a trotting horse. Paso Finos also walk, canter and gallop.

The ride to the ranch was full of adventures; it was my first experience at finding the giant *Anolis* lizard. This species was related to the ones sold at circuses for children to keep as pets, except that it was several times larger. As we were fording a shallow stream, one of these unusual specimens jumped off a branch and swam in the water ahead of our horses.

Life at the ranch was like a dream world for a city dweller from Cambridge. Bananas or plantains were served at every meal in one of three forms: raw, fried green, and fried ripened. Two times a day beef was also cooked in various forms.

The day after we arrived at the ranch there was great excitement because the task that day was to castrate mules. This was done by throwing the mules and removing the testes without anesthesia; the testes were all thrown to the same dog which gobbled them down immediately. What, I wondered, would this do to the endocrine system of that dog?

The most important excursion was to a giant cave near the seashore. We had to ride our horses through pastures filled with herds of oxen, and sometimes two of these giant bulls would be fighting. We wondered whether this would spook our horses or whether the bulls would attack, but all went well. At the shore near the entrance to the cave there were numerous land crabs which displayed colors of red, black and white, and were about the weight and size of a one-pound New England lobster.

The Cuban Boa, or *Majá de Santa Maria*, is the largest boa constrictor in the West Indies. It can reach up to 20 feet, but most are in the 10–12 foot range. In his office at Harvard University, John Welsh had a beautiful snakeskin of the Majá, and this of course made me determined to find one while in the field in Cuba. The Majá is known to inhabit the entrance of bat caves, and since this particular cave narrowed to a funnel entrance, it was reasonable that we would find a young Majá there, and we did. This was his station—when the bats flew out at night he could strike and seize his supper. Fortunately, I had brought a sack and we were able to place a very angry young boa constrictor in the sack; he was measured later at five feet. Eventually I brought it back to Cambridge as a live specimen.

This bat cave was totally different from any of the others. It contained a small free-tailed species that was quite different from the one I was studying at the station laboratory. I took a sample of live bats back to the laboratory to assess their stage in the reproductive cycle.

The cave itself was remarkable because of the layer of guano. In the caves I had been studying there was very little guano on the floor, presumably because the bats were fruit-eating. This was not the case with the free-tailed bats on the shore of the ocean. There was so much guano in the cave that it was very difficult to walk on it. We went outside and obtained a long dead branch which was certainly seven feet long, and managed to plunge that down into the guano without reaching the floor of the cave. Upon returning to the ranch I mentioned to the foreman through my interpreter that bags of this material would make a clear financial profit, if carted back to a plantation and used for fertilizer.

We spent one more night at the ranch, and the next day Felipe put the cage of live bats in his knapsack and I put the boa constrictor and two giant land crabs in mine. We started the homeward trek, which turned out to be a very different experience than our slow ride to the caves. It was an all-day ride, with living wildlife in packs on our backs. My beautiful mare became quite animated and went into her smooth, showy and fast *Paso largo* gait. With head held high, she charged toward home, giving me the ride of my life. People who know about gaited horses are aware that in this gait, it is possible to hold a full glass of water without spilling any of it. The mare was quite beautiful and as we passed through small villages, children and even adults would run over to watch her pass. It was a thrill; I could only think that if this stock horse were to be shown in Madison Square Garden, she would always get a blue ribbon.

After a few weeks at the laboratory preparing specimens, there was another memorable field trip.

John Welsh had admonished me to be sure to get into the Trinidad hills. This was easily arranged, but with different horses because the trail to the rainforest was too rough for the refined breeds. This is coffee country, and we shared the trail with caravans of mules carrying bags of coffee down from high up in the hills. All the produce of the coffee plantations was taken out of here on the backs of mules. Most of the time, the trail was barely wide enough for a mule carrying two sacks of coffee beans.

We spent the night at a coffee plantation halfway up the mountain. We slept on two couches in the living room after being served the most delicious coffee one could imagine. When I told the wife of the plantation manager that we bought coffee in our grocery stores where it might have been ground a month or more before, she was quite horrified. It seems that their way of making coffee was to keep the green beans in a tight tin container, and to roast beans and put them back in the air-tight tin. Then the roast beans are ground just before use and put in a cloth bag. Boiling sugar water is poured through the bag and finally into a cup. The result is the finest coffee one can imagine.

The next morning before breakfast I had one of my most memorable and enjoyable experiences in field work. I walked up a trail away from the coffee plantation into the rainforest and found a high cliff festooned with tropical plants. As I sat on the edge of the cliff, I realized that the branches of the trees in the valley below had intertwined and filled in so that their tops were like a carpet at my feet.

A most delightful concerto drifted up from the valley below. It came from about six species of Cuban birds; one has the beautiful name Peerless Solitaire; another was the White-crowned pigeon. I have never forgotten that symphony of birdsong. My eyes were also treated to a colorful vision as two spectacular flocks of birds, some parrots and others parakeets, flew back and forth just at my line of sight, all calling or vocalizing loudly; surprisingly those parakeets measured about the same length as the parrots, but had a more slender body.

The next day we continued up to the mountaintop and spent the night in a metal shack. I took pictures of the men who were living in the shack, much to their delight. They cooked some meals for us. I told them we were especially interested in checking up on the abundance of the Majá. They were delighted to contribute their tales of their experiences with this very large snake. But none of them had seen any in the vicinity of this ridge. Finally, we had to go back to sea level and joined one of the mule trains heading down the mountain.

We were there before the time of Castro and the revolution (which started in 1953 with an attack on the Moncada Barracks in Santiago). Later, as history was being made, I often thought about this remote group of men, wondering if they might have been one nucleus of very early rebel forces.

Next I would fly to Haiti from Santiago. The most eastern of Cuban cities, Santiago is near Guantanamo Bay, a parcel of Cuba on permanent lease to the United States. The atmosphere at that time was one transported from old Spain with sidewalk minstrels strumming guitars and singing ballads under the moonlight. This impression of old Spain was heightened by buildings and residences built of limestone in the Spanish style. This was the scene of Hemingway's *Old Man and the Sea*. In retrospect, because I wandered down to the fishing docks, I like to believe I met the "old man" himself.

I was also in Santiago to meet up with my friend Bryant Prescott from New Bedford. When I'd told Bry about my Haiti expedition, he had told me he would be in the Caribbean visiting friends in Jamaica, and asked if he could join me before he went back to the States. Of course I said yes, and thus, in the same way I had traveled to Cuba with Eric, I would also have a companion for a short time in Haiti. Bry met me in Santiago, and we flew together to Haiti.

## Haitian Adventure

I had expected to be a lone explorer on the expedition into the Haitian mountains, but Bryant wanted to accompany me on a trial expedition to the base of the Massif de la Salle, where I would gather specimens for Phil Darlington. We spent one night in a village at about 4,000 feet, and the next morning, with our sleeping bags and my collecting gear, we set off. I must say, we had quite an exciting time.

We did not have a guide as we packed into the foothills of the plateau. I was busy collecting on the way, and the Anolis lizards that I preserved for the Smithsonian Institute, were later to change the classification of that large group of lizards on Hispaniola. When a large green snake went across the path, I picked it up for the museum at Harvard. Some Haitians were approaching, and when they saw me holding the snake they fled back down the trail with their hands in the air. Interestingly, Bry and I had read some articles and books about the island and its Voodoo religion, and I suspected that their alarm was perhaps because the snake was considered sacred. The information we'd read mentioned two key figures in that belief system, the Mamma-loi and the Pappa-loi. Apparently Mamma-loi can incarnate as a green snake, and if the article was at all accurate, to a follower of this religion, it must have seemed that I was capturing the Mamma-loi.

**Along the trail in Haiti.**

We finally climbed up a rainforest-like ravine with a rushing stream in the middle. We camped by a large Banyan tree, and unrolled our sleeping bags on the rocky ground, avoiding several fairly large stones. As I climbed around in the Banyan tree, with its aerial roots and lattice-like trunk, I did catch some unusual *Anolis* lizards, but what was most interestingly was the giant tree frog that I caught.

Later I learned that this giant tree frog is appropriately named "Hyla vasta" *(Osteopilus vastus)*. It was about the size of the largest American bullfrogs. I decided I would take the Hyla vasta back alive because it seemed valuable and I wanted to photograph it.

I should mention that my dear friend Bry had a temper. He had taken photographs of some Haitians who had passed our camp, and they approached him requesting money since he had taken their picture. He pretended to be angry, and said, "No, no, no! I don't give money for taking pictures." They seemed slightly offended, but they went away.

That night it was the time of Easter celebrations. At dusk, we heard drums throughout the hills; it sounded as though groups were sending drummed messages back and forth. More importantly to us, some of the drumbeats seemed to be heading our way.

It was very eerie to hear the drums getting closer and closer, and it reminded us that we were, after all, white intruders in the hills of Haiti. And sure enough, right to our campsite, right up to our sleeping bags, came people in costume beating drums and dancing. There was one who we guessed must be a medicine man making gesticulations and pointing at us as we stood stoically and smiled. He had numerous artifacts, possibly sacred objects, in his hands, and these he pointed at us. We presume he had put a bad spell upon us for the present and perhaps for the distant future because Bry had angrily refused to give them money for a picture. We supposed they wanted money, but it really didn't seem wise to encourage them at that particular moment! Then they turned around and left, beating their drums back down the long trail—and we could hear it for a long time off in the distance.

We went to bed and started to get a good night's sleep on the rocky campsite, but at three in the morning they all returned beating their drums and acting like they had before. We got a double dose of the medicine man's "sacred incantation" as we repeated our stance of smiling stoically again, and not giving them money. Then again, off they went through the valley still beating their drums.

The next day, carrying my lizards, the green snake and the Hyla Vasta, we returned over many trails to our hotel in the mountain village. There I tied some string about the valuable lizard, which was still alive, and the Hyla Vasta. I could take the lizards into the sun to take their picture but I decided to photograph the very valuable large tree frog in my hotel room. I released it to see what would happen and it gave a giant leap to the ceiling of the room and held on upside down. I pulled it down and took it to a spot of sunlight in the hotel room, photographed it, and then very sadly euthanized it for the Museum of Comparative Zoology in Cambridge.

Bry and I returned to the hotel in Port-au-Prince, and he flew off to New Bedford, while I continued my stay.

When I had begun making my plans for Haiti, while in Boston I had met a charming and handsome gentleman who turned out to be the manager for the Cowley Fathers' monastery on the Charles River in Cambridge. These priests are an order of former Anglican monks who became part of the Episcopal Church when they arrived in North America from England in 1870. When this kind gentleman, Rupert Jenkins, heard that I was going to Port-au-Prince, he told me he would also be there at that time, and offered to introduce me to the Episcopal Bishop for Haiti, also a member of the Cowley Fathers.

After Bry left, I contacted my friend Jenkins, who immediately introduced me to the Bishop. I was then invited to travel into the rainforest at sea level to attend a christening for a group of converts to the Episcopal Church. Bishop Burton was the host, and the next morning he led the way on horseback over a difficult trail through the rainforest.

I had been provided a horse by the entourage and noted that the cinch that held on the saddle appeared to have been fastened with some stalks of grass. Sure enough, when we forded the stream, my horse insisted on drinking, but since the other horses were not drinking and I worried that he was too hot to drink, I tried to pull his head up. At that, the grass holding the cinch pulled apart and the saddle slid down over the head of the horse and, as it was a fairly rapid-running stream, I started down the watercourse still riding the saddle! The Bishop, the other priests, and the entire entourage decided it was the funniest sight they had ever seen. We soon had the saddle back on the horse and proceeded.

We came to a church after an extended morning in the rainforest. There were many shining black faces; the youths were mostly 10–14 years old, and all wore white. The ceremony was long and beautiful as they were brought into the Episcopal Church. After that we turned around and rode back, arriving in Port-au-Prince after dark, this time without any unexpected rides down the river.

It was now time for me to plan the difficult expedition to cover the whole distance of the plateau on the *Massif de la Selle*. Word had spread, and from a village a few miles up the mountain a nice-looking Haitian man named Pico appeared and offered to join the expedition. Much to my delight, I found that he was the same man who had been the guide for Phil Darlington, the professor who had persuaded me to take this trip on the plateau. I knew a bit about what to expect because I had read a description of the area by Ivan Sanderson in his book, *Caribbean Treasure*.

I had to go to the neighboring town of Pétionville to pick up Pico. We had to travel by hired car. Our first destination was the small village at about 4,000 feet and the hotel where I had taken pictures of the captive Hyla Vasta. After we arrived there, I stayed at the hotel and Pico stayed with friends.

The next day we were off with one large knapsack containing all of his belongings and mine, including a sleeping bag, a small bottle of ether and a preservative for my specimens. Pico occasionally carried the pack on his head but preferred to wear it on his back with his arms through the straps like an American hiker; I was paying him twenty cents a day.

**My Haitian guide, Pico, posing with our pack.**

It wasn't long before he subcontracted that task to a boy who looked to be about fourteen years old for five cents a day. This boy balanced the knapsack on his head as he walked cheerfully up the mountain trails.

We camped at the base of the trail leading up the massif, and I took pictures of the stream of individuals coming down the very steep trail. They were going to Jacmel, a village on Haiti's southern coast. There was also an opposing stream of people moving in the other direction. Despite the tortuous trail, some men and women carried children on their backs. Others drove donkeys in an incredibly cruel fashion, beating them as they stumbled down the rocky trail. I never could learn why these families were driving the donkeys, whether to use them as a beast of burden when they got to their destinations or whether the donkeys were to be sold.

As we traveled these trails, I was filling my notebook with some Haitian dialect names. My guide received some of his nourishment as we walked along the trail from a bottle, which he allowed me to smell. It was obviously quite fermented and alcoholic. The name he gave me for the material in the bottle sounded incredibly close to something gardeners use as they maintain their plants; as Pico said it, the name sounded like "*Composssti.*" (I am obviously referring to compost piles from garbage and other organic waste that can be used to enrich the soil.)

Throughout the stay in Haiti, I was collecting insects for Phil Darlington. Although he had once taken this same trip, I assumed I might collect some insects that Phil had missed. Pico had been well trained in insect collecting by Phil; indeed, he was so eager to catch insects that he displayed an amazing tolerance to bee stings. There were many flowers along the trail; the most common, surprisingly, were the parent stock of the beautiful red Poinsettias that in North America we buy at Christmastime. Pico went to any blossoms where bees were abuzz, and snatched the tiny pollinators from the flower, put them in our vials, and then calmly pulled the stinger out of his hand. Perhaps the *Composssti* was responsible for his tolerance to the stings?

Competing with the travelers coming down the steep, rocky trail, we finally made it to the top of the plateau. That first night, we put our sleeping gear on the ground close to the trail because darkness had fallen.

A biologist is fortunate to have some experiences that stand out during the rest of his career, and while we lay in our sleeping bags, I had such an experience. I heard a sound from under the rocks around my pillow that can only be described by imagining a xylophone made up of very small, but different silver pipes. Now imagine a little metal hammer hitting up the scale on these small pipes. Out from under many nearby rocks this beautiful tune erupted. Whenever I shined a flashlight in the direction of the sound it stopped. The next morning I managed to capture some of the singers (they turned out to be small frogs) for my home museum.

**Haitian Tree Fern, one of the oldest living plants.**

We continued our trek, crossing the length of the entire plateau. We slept on the mountain for two nights, and kept busy in the daytime collecting insects, but there were very few lizards. Here I found and photographed my first tree ferns, the oldest living plants except possibly the Cycads. These tree ferns were found in very restricted places, mostly sink holes made of rock that looked, from a distance, almost like marble.

Surprisingly, the most interesting event we saw as we walked to the end of the plateau was several horsemen who were trying to break some local horses. Haiti's horses are descended from the Barb horses brought to Haiti by the Spanish in the $15^{th}$ century. Aside from Paso Fino and other well-bred horses owned by ranches and other big outfits that can feed them adequately, the equines seen in the small villages and jungle were small, wiry and undoubtedly tough, having had to survive in a country plagued for so long by poverty and malnourishment. These particular beasts were pony-sized, light brown, and had a surprising amount of spunk.

As we descended the steep trail at the end of the high plateau, I remembered a story Phil Darlington had told me about another time going down this same path when Pico had been guiding a professor, and the man had fallen. Therefore, I was not surprised when Pico practically tried to lift me down parts of the trail down from the plateau. In fact, his care was so evident it was the dominant experience for me in going down the homeward trail.

**Plateau trail, Haiti.**

After getting to the foothills, we stayed at a very small village in a precarious location on a hillside. Here I made a study skin of a red tailed hawk that had just been shot by a soldier who was on patrol. We were befriended by a young man named Roger Victor, who was an agricultural agent. He spoke excellent English and allowed us to sleep in his only bedroom. The bed that I was on was at such a slant, I kept rolling off. Lucky for him, Pico was in another part of the room on the floor.

That night was characterized by the abundance of rats. First I heard some of them fighting on the tin roof. Then I happened to turn on my flashlight and saw two of them gnawing on my knapsack beside me. And finally, I woke to a rat on my face—it was sniffing my lips! I sat awake with the flashlight on the rest of the night, thinking fondly of the Cuban Majá, the snake that eats rats. I must have dozed anyway, because at daybreak I was awakened by the calls of a very large terrestrial bird, obviously a relative of the Rail family. I had seen only one or two in Cuba but it appeared to be the same species here on a small mountain in Haiti.

Our friend Roger Victor had so enjoyed talking to us that he presented me with a published story by Victor Hugo and signed the front of it with his name. Just as we were leaving to walk back to the hotel at lower altitude, the soldier in uniform who had shot the hawk joined our party. As usual, I was amazed at the lack of footgear on Pico and the small boy who was still carrying the pack balanced on his head. They were barefooted on this rocky trail, and in all of this expedition, the only footgear I saw on any of the local residents was on the soldier who had now joined our party.

Off we went once again over the stony trails to our goal: the place I began to think of as the "Hotel Hyla Vasta." This was a full day's march. My two companions were eager to get home and we walked fast up and down the small hills. Now for the first time I found that Pico was so impatient that he would not stop to empty his bladder. When this needed to be done he merely watered the trail as we walked along.

The only other observation I can make concerning the hundreds of local inhabitants we had met was that the women were all topless. This surprised me because I would have thought that they would cover their upper torso as protection against insects and branches along the trail—if for no other purpose. I am sure that the reason for this lack of clothing was to assist in temperature regulation, because even at the top of the 6,000-foot plateau, the tropical sun beat down and it was uncomfortably warm, as it also was in the foothills we had just traversed.

At the next village high on the hill, we again stopped for the night. The soldier who was walking with us brought some sick children to me. I did have some first-aid ointments, which I applied. Now the soldier was striding along with us on the trail to go back to that small hotel where I had photographed the Hyla Vasta on the ceiling of my room. Occasionally, the officer developed a mild dictatorial style toward the natives if he didn't like something they were doing, or he didn't want an American observing. When we arrived at our small hotel, we were still approximately 4,000 feet above sea level.

I had initially arrived there by taxi. The next day for my trip back to Port-au-Prince there were no taxis around and no telephone on which to call one. The officer, who spoke very good English, insisted above my protests that I ride his horse back to Port-au-Prince, which was where he lived. His horse had been maintained in the village and when he brought it forth it appeared to be a fine animal possibly left by the marines when they had been in Port-au-Prince for security. At any rate, in spite of my reluctance, I was soon astride the horse and the officer walked behind while we descended the steep trail. I very much enjoyed traveling in the same direction as the stream of villagers going down to the market. Some of them were astride donkeys, but I didn't see any other horses.

**Helpful soldier and his wife.**

When we arrived at the man's house he asked, "Now will you take my picture with my wife?" Of course I agreed, despite having a previous appointment with Bishop Burton. I waited for what seemed a long time while I heard outdoor showers being used. Finally the soldier appeared in his Sunday suit, not his uniform, and his wife in what was probably her very best dress with her hair nicely fixed. I took a number of pictures of them, and was on my way. Of course I sent them enlargements when I returned to Cambridge. I know I had the right name and address but never heard whether the pictures arrived in his house. I spent one restful night at the City Hotel in Port-au-Prince.

Soon it was time to pack up the specimens from Haiti and to load the plane for my return to Santiago, Cuba. I eventually sent the Haiti specimens to the Agassi Museum and the Smithsonian Institute.

Back in Cuba, I spent another delightful night in a small hotel in Santiago before the day-long bus ride down to Cienfuegos, on Cuba's southern shore. In those days a bus trip was eventful because similar buses tried to race each other on the two-lane highways. Fortunately there were few curves and valleys, which allowed the races to be exciting but without serious incident.

The last third of the trip was after dark. I was dozing away and happened to wake up as a very large white object came into the open window at the front of the bus. This object came down the aisle and landed on my knee. I recognized it as a Cuban Barn Owl and seized it, causing it to sink its talons into my knee and thigh. I wanted to photograph it and was able to wrestle it into the large camera bag I carried. We had seen several of these raptors floating like ghosts down some of the long caves in which we captured bats. Upon arriving at the station I managed to attach leather thongs to its legs to restrain this beautiful bird. After I had photographed it back at the laboratory, I carried it by horseback to the nearest cave and released it.

Before I left the West Indies, there was one last cave to be explored in Cuba. Felipe and I traveled there by horseback, and because there were no accommodations nearby, we camped out, sleeping in our clothing for three days.

## Back to Cambridge

Leaving right after that last primitive camping trip in Cuba, I transported all the Cuban specimens to the Department of Biology at Harvard, and then moved into a Cambridge boarding house run by Tina McDougal.

There was a curious incident soon afterwards. Although I certainly had the appropriate number of baths at the boarding house upon returning from the West Indies, after a few days I began to itch, especially on the chest and armpits. Student health determined that I had a bad case of "crab lice" due to sleeping in my clothes while camping during the last horseback trip in Cuba. A medical entomologist at the Harvard School of Medicine heard about this and telephoned saying he wanted to make slides for the medical students of my particular brand of lice. I presume it could be possible that now, fifty years later, some Harvard medical students are still learning to identify these parasites from my own personal lice.

Not long after I returned to Cambridge, there was a photography competition for the entire Department of Biology at Harvard. Graduate school had included long hours in the darkroom, and I had learned how to develop film and make prints on different kinds of photographic paper. Although I had many pictures from Cuba and Haiti, the one with the best definition was of the Cuban Barn Owl that had flown down the aisle of a bus to land on my leg. That photograph took first place in the contest.

I spent the rest of the summer making histological slides of the material collected in Cuba, and writing up the story of reproduction in the bats at that latitude. My friend, histologist Donald Fawcett, eagerly studied my specimens of the placenta of the *Artibeus* bat. He found the placenta to be quite unusual and published the results.

I sent the ovaries from the *Anolis* lizards to Hermann Rahn, who had by then moved out west to teach zoology at the University of Wyoming. He did not discover lutin-producing cells in the ovaries of the *Anolis* lizards from Haiti. But Hermann was also studying rattlesnake reproduction at high altitude; he observed that female rattlers living above 6,000 feet where summers are short have a 2-yr reproductive cycle, while the same species living at lower altitudes with longer, warmer summers reproduce each year.

To continue my own research, fortunately I received the utmost cooperation back at the Institute in Cuba from David Sturrock and Frank Walsingham. They arranged for one trip per month to our major study area, which was a large limestone cave. The only way it could be studied by the guide who had been with me on earlier trips was to bring back a sample of the bats. He not only obtained the bats, but also dissected out the reproductive tracts, put them in alcohol, and sent them to me to be measured.

The results were already evident from examination of the specimens: instead of having a breeding season scattered all over the year, as might be expected near the equator, the *Artibeus* bat had a very specific breeding season which was limited to a few months centered on the month of March.

# *Chapter Six*

## *World War II: Harvard Fatigue Laboratory*
**1940–1947**

### Teaching at St. Mark's School, and a New Family

THE FINANCIAL EFFECT had again taken hold, so before I could continue my studies towards a PhD, I returned to teaching biology—this time at St. Mark's Preparatory School in Southborough, MA, just thirty miles west of Boston.

This new chapter in my life also gave me an opportunity to ask for the hand in marriage of Dorothy Dodge, the young woman I met while working for the Prescott family during summer breaks in graduate school. We were married about a year later in Lydia (Prescott) and Gordon's new house in Cambridge. Dorothy and her four-year-old son, Christopher, joined me in Southborough.

Soon we had a new addition to the family—of the four-legged variety. Dorothy's mother gave us a beautiful Boxer puppy. He was brown with a few white markings and had just been imported from Germany; we were told he was very well bred, as his father was Zunftig Von Dom, a famous foundation sire. We named the little guy Hans.

That dog was devoted to the family and especially to me because I took him for walks and sometimes even to the lab at St. Mark's. As we crossed wide-open areas on campus he was like a tame eagle, running far away, circling, and then landing back at my feet. He was a real athlete, all legs.

Within a few years, we celebrated a much more important arrival—our daughter, Victoria Elisabeth, was born July 20, 1942. Christopher was delighted that she arrived the day before his sixth birthday. With these back-to-back birthdays, as the children grew, July was welcomed as the birthday month.

Even as a baby, Victoria liked to "help" me feed Hans. I remember one afternoon when Dorothy had left me in charge of the kids. I must have dozed as I sat on the carpet where Vic was propped up playing with some toys. I awoke from my nap realizing that our baby had crawled all the way to the other room: there she was in the kitchen, precariously balanced on hands and knees, face in the dog's food dish as she lapped, or pretended to lap up his food. I rushed to scoop her up, but Hans wasn't the least concerned.

Meanwhile, I enjoyed teaching at St. Mark's and was setting up a biology laboratory for the students. Don Griffin flew in to visit me and show me his new airplane, which he used in his research on bird migration.

At an institution like that, you become very good friends with other faculty members; I spent a good deal of the time with Richard (Dick) Eberhart, a published poet, and Hale Very, an accomplished pianist. Hale's grand piano sat outside my laboratory and I often watched him practicing Chopin—so deeply did he feel the music that his face became enraptured.

## War

Peaceful though it was in Southborough, across the ocean a war raged. The Nazis had invaded Poland in September 1939, and it wasn't long before most of Europe was at war. The United States was for the moment content to sit on the sidelines, but I was eligible for the draft and had to consider the possibility that the US would also become involved. I wanted to serve, but not to kill, so I took a non-combat assignment in the Medical Corps of the National Guard in nearby Framingham. My lifeguard training brought me the rank of Sargent. The next few years were anxious ones, as my family and I did not know if and when the Guard might be called out.

On December 8, 1941, Pearl Harbor was bombed, and the United States was at war, fighting Japan, Germany and Italy. I continued to teach while wondering how I could serve my country and support my family. I was offered a commission in the Army as a gunnery officer on a transport ship, but *The Financial Effect* excused me from taking that particular job because the salary was not adequate to sustain my growing family.

### *Hans Gets Drafted*

Kind as he was to each person in our family, Hans started to show unfriendliness to strangers. I attribute this change in his temperament to the actions of two inexperienced people who reacted to him in a way that put him on guard. The first was a teenage babysitter. We were all in the living room, and when Hans trotted in she crouched down and cried, "Oh, he looks vicious!" I suppose he was bothered by the way she was acting, so he crouched as well, and he did look rather fierce.

The second person who did this was a janitor at work who reacted the same way, cowering and saying that Hans looked vicious. Again, Hans also crouched, and this time he began to growl. I quickly took charge of the dog and nothing further occurred, but after that, he began to growl whenever he was unsure of anyone.

I grew worried that he might bite somebody, and with a heavy heart, started to look for a different kind of home for Hans. Around that time, the US Army put out a call for dogs to serve with our troops. Thus, Hans joined the Canine Corps. We were sorry to see him go, but the Army was happy to get such a fine looking Boxer, and gave him a job patrolling a military airport where his guarding instinct could be put to good use. The Canine Corps promised us that he would be sent home after the war.

Life went on, but the war was always in the forefront of everyone's mind. One day in 1943, Hale Very, Dick Eberhart and I were having lunch in Cambridge with another close friend of mine, Gordon Thayer. Gordon had joined the Merchant Marine and had been across the Atlantic several times. As we discussed how we could contribute to the war effort, Gordon encouraged us to sign up for the Merchant Marine. At least the income would be adequate to support a family of four.

The Merchant Marine, although a commercial entity, was integral to the war; it built transport ships and conveyed troops and freight for the military. Joining the Merchant Marine would satisfy both the draft and my financial situation, and thus I resigned from both the National Guard and St. Mark's and signed on.

After passing all investigative procedures, I was assigned to a ship and had been on the dock several times awaiting orders. Because of the danger from enemy submarines, ship departure dates were highly secret—after reporting for duty with our packed bags we would often be sent home to wait. Meanwhile, my musician friend Hale had been trained as a construction sergeant in the Army and received orders to cross the Atlantic on a troop transport. Hale and I hoped to be assigned to the same ship.

### *Altered Plans*

While waiting for my ship to sail from Boston Harbor, I received a phone call that would change my course not only for the next several years, but also for my entire career.

The caller was Donald Griffin, the brilliant biologist who was also my close friend and colleague. Don was with the Fatigue Laboratory at Harvard, working with other physiologists on wartime research. Much of their effort was directed toward developing and testing equipment for soldiers in cold and extreme environments.

"You can't go into the Merchant Marine, Ed," he said, "I have a job for you here at the Fatigue Lab. We need more people trained in science, and you also have experience with winter camping," he added, referring to all my trips to Mt. Washington during eight university vacations.

I accepted eagerly, as this would fulfill both my military and family obligations, and moreover, would allow me to return to something I loved—research. I resigned from the Merchant Marine and began to look for a house in Cambridge.

That same week, Hale Very boarded a transport and shipped out of Boston Harbor. It was the height of the Battle of the Atlantic, and tragically, my warm, gifted friend would never make it to Europe; his ship was sunk by a German submarine and there were no survivors.

Years later, I still remember in detail that lunch with Dick, Hale, and Gordon, talking about the situation of the war and wondering what our relationship was to the larger scene. I never heard how Dick Eberhart spent the war years, but after it ended, he continued developing his fine reputation as a poet, producing work until the time of his death in 2006. Gordon Thayer stayed in the Merchant Marine, and was paymaster on various ships.

My plan for the war had been the same as theirs: to serve on one of the more than 200 ships crossing the Atlantic. Half of these were sunk. I'm tempted to believe that my early love of winter camping may have saved my life by redirecting me from a ship to a military laboratory in Cambridge.

## On the Staff of the Fatigue Laboratory: 1943–47

I spent the next four years at the Harvard Fatigue Laboratory, and those years alone could provide enough material for an entire book. At least one book *has* been published about this famous laboratory: *The Harvard Fatigue Laboratory, Its History and Contributions* (1973), written by my friends, Steven and Elizabeth Horvath. Their book provides an excellent description of the Laboratory's development and early years, yet devotes only nine pages to its contributions during and just after the war. According to the Horvaths, the Fatigue Laboratory sent some 180 technical reports to military offices of the US, British, and Canadian governments throughout that period. "To enumerate these studies," they wrote, "would be exhaustive."

But it was an exciting and historic time to be at the Fatigue Laboratory, and I am happy to be able to record some of the testing and design we undertook to improve the lives of our troops and to support the war.

The Harvard Fatigue Laboratory had been created by the great biochemist L.J. Henderson along with David Bruce Dill in 1927. It had been established at the Harvard Business School and given the name *Fatigue Laboratory* because its original mission had been to study the physiology of fatigue to determine its effect on industrial workers. Many of its large-scale projects consisted of studying the effects of heat and cold on individuals, and attempting to make them more comfortable during work or exercise in extreme environments.

An example of this was the building of the Hoover Dam in the Nevada desert: workers in the extreme heat were succumbing to heat exhaustion and heatstroke, and there had been 13 deaths (Dunar and McBride 1993). Bruce Dill and John Talbott went to the site and ran physiological and biochemical studies on themselves and healthy work crew volunteers in the heat. They recommended ensuring that the men had cooler conditions in which to sleep and recuperate after working all day in the heat, and also to add copious amounts of salt to the men's food.

**Lunch break at Fatigue Lab. *L to R, standing:* S.M. Horvath on the left, I'm on the right, *sitting:* W.H. Forbes, Sally Smith, H.D. Russell.**

The Fatigue Laboratory was one of the first exercise laboratories in the United States, and became the leading center in the US devoted to research on exercise physiology (Tipton 1998). There was a hot room that served as an "artificial desert" and could be run at temperatures up to 115°F, and a cold chamber that could be operated from 40°F down to -40°F. Both were equipped with treadmills and metabolic equipment to measure subjects' physiological responses, including metabolic rates at rest and during exercise.

At the time I joined, this laboratory had the only experimental high altitude chamber in the United States. It was installed in the cold room, and experiments were done on oxygen masks, the bends, electrically heated garments, and the physiological effects of suddenly breathing air containing low levels of oxygen. Our altitude chamber was a model instrument, but it was scarcely used while I was at the Fatigue Laboratory; there is one early technical report on its use in 1942. Very quickly during the war, new altitude chambers were constructed at various places throughout the United States, most sponsored, of course, by the Air Force.

During the war, the day-to-day director of the laboratory was William H. Forbes, but the person who had developed the interest in extreme environments and acted as director after L.J. Henderson had retired was Bruce Dill. Dill, however, had been commissioned in 1941, and left on temporary assignment to the Aeromedical Laboratory at Wright Field in Dayton, Ohio. After two years he was transferred to the Army Quartermaster Corps in Washington, DC, where he attained the rank of Colonel. Although at a distance during most of the war, Dill was able to stay involved and "help direct the program of the laboratory toward the solution of pressing military problems" (Dill 1966).

**Cold chamber staff. *L to R:* H.D. Russell, H.S. Belding, G.E. Folk. Harvard Business School is in background, with Fatigue Lab in basement.**

The expert on cold was Harwood "Woody" Belding, and the expert on heat was Robert E. Johnson. Working with Belding, I was to take charge of the cold chamber and, using between four and eight soldiers and myself as subjects, develop improved cold-weather clothing and equipment. It is important to note that at the Fatigue Laboratory, we were all subjects in any experiments we asked the soldiers to endure.

Three especially effective researchers, who also became my lifelong friends, were Robert Darling, Fred Sargent, and Robert Kark, all of whom had been at the Fatigue Laboratory since before the war. All three worked in both the hot and cold chambers in the Laboratory, although Bob Darling turned his activities solely to the cold room at about the time I arrived. Sargent, an expert on the physiology of aging, ran his own program, which also operated across all research areas.

### *Working in the Cold Room*

The Fatigue Laboratory's program to test clothing and equipment for use in cold climates was coordinated with a similar program at the Climatic Research Laboratory (CRL) in Lawrence, MA (later moved to Natick, MA). This facility was developed during the war and had been patterned on the Fatigue Laboratory; John Talbott became its head. The two laboratories shared a company of soldiers who volunteered as subjects.

The usual question from the Quartermaster Corp of the Army, the Air Force, or the Navy was to describe the insulation value of various garments whether dry or wet. The insulation might be wet because of immersion in fresh or salt water, from rain, or from perspiration originating from the skin soaking the insulation. The Fatigue Laboratory had developed a variety of techniques and calculations to answer these questions. The insulation value for each garment or group of clothing was usually expressed in terms of *clo* value.

One *clo* is about the insulation of a business suit for a person sitting in an environment of 70°F. A soldier remaining inactive at -40° would require an insulation of about 6 *clo*. Of course for various garments, hand gear, and especially foot gear, the *clo* value would have to be measured individually. These units could also be used to measure the insulation values of fur on animals.

**Tumpline: How to carry 220 lbs.**

This system and formula for measuring the *clo* value of various outdoor outfits had been developed a few years earlier by Alan Burton and colleagues Gagge and Bazett (1941).

Alan Burton was the great cold weather physiologist who worked in England and Canada during WWII. (Bazett and Burton each later served as President of the American Physiological Society.) Burton organized and led a cold weather school in Toronto, which Bob Darling, Bob Kark and I attended in preparation for teaching camping in the cold. We were amazed when the instructor showed us how to carry 220 pounds using a tumpline, which is a strap worn over the front of the head attached at each end to a pack or sling. This could come in handy for our soldiers on the fighting front, allowing them to carry wounded comrades while leaving their hands free to hold weapons.

There were other ways of measuring insulation: copper models of a typical human body and models of the hand and the foot were used to measure heat flow from the interior source of the models when they were dressed in different garments at different ambient temperatures. These results could be checked against actual heat flow from human subjects in the cold room exposed to various temperatures.

There were three techniques used for measurement during the recording of insulation of various garments. The first and major measurement was recording skin temperature of ten parts of the body; these locations, known as Belding Points, have become standardized in the literature (Folk 1998, p. 445). Because the source of heat was more important in some parts of the body than in others, these measurements were weighted as far as the surface area for that particular region was concerned. The second measurement was the energy metabolism during the wearing of garments either while completely stationary or while walking on the treadmill in the cold room. The third measurements were made with a radiometer to record radiation from the surface of the garments.

I will describe the kinds of experiments we did on human subjects; these were usually soldiers, Fatigue Laboratory staff, or several of each. As this was in the 1940s, the soldiers were always male. We used a questionnaire to find out the health of the human subjects on that particular day. Then we taped thermocouples on the skin to measure skin temperature at the ten standardized body points. We had the subject(s) dress in the experimental uniform being studied. Next we would take them into the cold room, seat them as comfortably as possible considering the unpleasant environment (pilots had to sit in the cold for long periods), and give them a mouthpiece and a clip on the nose. The mouthpiece was used for ventilation, and allowed us to measure the metabolic rate. Metabolism is measured by breathing oxygen out of a tank, where a pen records on paper the $O_2$ used by each breath. When exhaling, the air from the lungs passes over a $CO_2$ absorber.

One day Donald Griffin decided he needed to do an extended experiment on metabolism in the cold. He asked me to dress appropriately for -40°F and to breathe out of this metabolic equipment in the cold room, which had been lowered to that temperature, from 8:00 AM until 8:00 PM. If the experiment with me succeeded, then he would carry it out on one of the soldiers assigned to the laboratory. By the seventh and eighth hour of breathing oxygen at -40°F, I began to wonder what dire effects might occur. There were none, however.

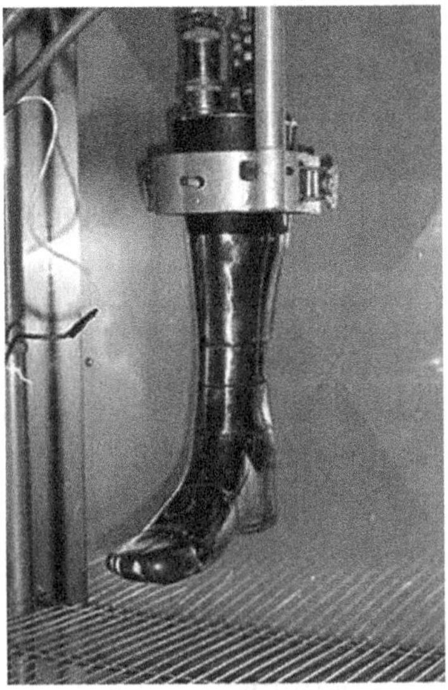

**Artificial heated foot model used to measure insulation. (Photo: Thomas L. Endrusick.)**

We had read stories about military barracks in the cold regions which caught on fire while the occupants were sleeping. I was given the assignment of determining what clothing should be worn when escaping a fire at air temperature -40°F, and to determine how fast to run to maintain body temperature when inadequately clothed in such an extreme environment. Two of the soldiers joined me after I first tried this test. After several changes of clothing we finally decided that the long woolen underwear issued to the troops in the cold regions would be typical of how soldiers would be clothed when they ran out of a burning barracks. I did wear gloves and tennis shoes and had a scarf around my face. We began at a moderate speed and the balance between heat loss and heat gain proved adequate. We kept increasing the speed and I found I was perfectly comfortable to run at 7 mph in just woolen underwear at -40° for at least 30 minutes. Afterwards, the two soldiers repeated the process and soon we had enough data to write a report to the Army Quartermaster General and the Air Force as to what was reasonable emergency garb to wear when escaping from a fire in extreme cold.

I recall an interesting experiment done with the soldier subjects concerned with the encumbering found when heavy garments are worn in the deep snow and the severe cold. Of course our arctic uniforms, good for -40°F, were quite heavy. I wondered how much extra energy it took to wear these uniforms and believed I could fractionate this amount of extra energy due to the weight of the clothing and its encumbering effect. We were able to fractionate the contribution of the two factors, and I named the possible contribution of the encumbering clothing, "The Hobbling Effect" (Belding et al, 1947a, 1947b).

Another intriguing problem assigned to me was the movement of air within arctic clothing. The Inuit had solved the problem of trapped moisture several thousand years ago. We had studied their clothing and diagrammed it for the purposes of developing army equipment. Therefore we could focus down on the effect of trapped air within not only arctic clothing but even within the fur of a sled dog. How much does insulation change when a person is resting in an arctic uniform and when there is action because the person is walking, perhaps in deep snow? Much to my delight I was asked to invent an instrument to measure the insulation when the uniform covering an arm was ventilated and moved or when it was quiescent. In the Laboratory's machine shop, I worked with a machinist to design and make a copper artificial arm that could be moved and pumped and studied with different kinds of insulation. I was also able to put the hide and fur of a sled dog on the artificial arm to determine the insulation of a sled dog while being quiet and again while running. Not surprisingly, I found that the insulation decreases when it is moving.

Although World War II aircraft were not pressurized or heated, some flew at heights of 25,000 to 30,000 ft—as high as Mt Everest. Harwood Belding was supervising the work of developing the electrically heated suits for the crews of the B-17 *Flying Fortresses,* the high-flying heavy bombers that were sent out on missions over Berlin. This gave us a lot of testing to do on ourselves to figure out the optimum way to distribute and calibrate the electrical heat over hands, arms, legs and torso at -40°F, the environment experienced by our aviators on those flights. Thus, we did many tests with experimentally heated suits on how to best provide heat, and what was the optimum amount of heat needed by the various parts of the body at that temperature.

**Donald Griffin and the equipment he designed to distribute the heat in gloves to be used at -40°F. (Griffin and Folk 1944.)**

One of the tests was designed by Donald Griffin, and I was the subject. I dressed in the experimental electrically heated suit and entered the cold room, which had been lowered to -40°F. Don then ran electric wires from the different points in the suit to a row of rheostats. His job was to raise or lower the heat in the various parts of the suit and my job was to tell him if I was too cold or too hot in that area. When we started out I told him, "I am too cold all over!" Soon I was telling him, "My arm is too cold…my foot is too warm…my backside is too cold…my elbow is now too warm…" and Don would record figures, adjust the rheostats, and in this way we could ultimately determine the most comfortable temperatures for which parts of the body when at temperatures of -40°F. We felt inspired by the knowledge that our aviators would be wearing suits like these, heated to the temperatures determined by us, in their life and death missions in the war.

As always, after any experiment commissioned by the government, we wrote reports for Col. Bruce Dill or other appropriate official in the Army Quartermaster's Office. In this case, our findings were also sent to General Electric, where the suits were designed and produced using our figures.

### *Cold Injuries and the Battle of the Bulge*

Throughout the time I was on the staff of the Fatigue Laboratory, much of our effort was devoted to preventing cold injuries, especially of the feet, the hands and the face. These injuries differed in respect to several factors: the conditions under which they occur, the rapidity of their development, the severity of the injury, and their characteristic features (Giesbrecht and Wilkerson 2006).

Our Armed Forces assigned to the northern regions had injuries due to dry cold. They encountered two kinds of injuries: true frostbite, in which the tissues actually freeze and form ice crystals, and milder injuries in which the tissues are not actually frozen. Of course, frostbite can develop with remarkable rapidity, but in the Fatigue Lab we were more interested in those cases which developed gradually over a period of days. This type of frostbite is caused by direct contact between the exposed part and a cold object—such as delicate instruments in a cockpit when the air temperature is -40°F, often the case during high-altitude flying.

In another group of injuries, wetness is as important as the cold itself. "Trench foot" is the name given to severe cold injury occurring over periods of days in ambient temperatures slightly above freezing combined with wet or moist conditions; "immersion foot" describes the injury suffered when feet are exposed to cold water for prolonged periods of time. In these injuries, the tissues do not freeze, but the resulting damage at times appears identical to that caused by frostbite. A consequence is *necrosis*, or death of tissue, which damages somatic nerves that provide sensation or motion. Even temporary paralysis of important nerves results in difficulty walking, sometimes described as a "foot-drop gait." The task at the Fatigue Laboratory was to design equipment to prevent such damage, and to try to insist upon its use.

Because of the Fatigue Laboratory's earlier tests, we knew that the standard leather military boot was inadequate for cold weather service. Such boots are not impermeable to moisture, and do not permit extra pairs of socks to be worn. If soldiers were serving in trenches or foxholes under snow or wet conditions, they must be dressed with three pairs of wool socks on their feet. Therefore, the Fatigue Laboratory's tests included a boot with a higher last, allowing for more socks to be worn by the soldier. After testing, the Fatigue Lab recommended the LL Bean shoepac, a sturdy boot with a rubber sole and sides, leather upper that laced around the ankle, and room for three pairs of socks.

Harwood Belding dressed a mannequin in this uniform, including footwear, and presented it to Colonel Georges Doriot, an officer in the Quartermaster Corps. Col. Doriot responded favorably to the advice, and discussed with numerous generals the advisability of changing the cold weather foot gear of the infantry being sent to Europe. Unfortunately, some of the generals insisted, "Our boys must march on leather!" To show his dismay and disagreement with the decision, Col. Doriot responded by installing the Belding mannequin, complete with shoepac and three pairs of socks, in a glass case in the Quartermasters' office, adding a note stating that this uniform was recommended in early 1942—two years before the Battle of the Bulge.

It is well known that during the Battle of the Bulge in the winter of 1944–45, our soldiers resisting the attacks of the Germans near the Dutch border were inadequately clothed. At times the air temperature was -30°F. They spent prolonged periods in trenches and foxholes where their feet were exposed to cold and moisture; temperatures were often around freezing, and trenches were at times flooded. For a long part of the battle, their footgear consisted of leather boots designed to hold one pair of wool socks. Later on during the battle, some old-fashioned buckle overshoes were issued, but one can imagine the problem of trying to get a wet leather army boot into a rubber overshoe. Additionally, the boots were difficult to remove rapidly in order to replace wet socks.

Because of this inadequate insulation, coupled with cold, wet, cramped and often immobile conditions, many soldiers ended up with trench foot. The first symptoms would be stinging or tingling in the feet, and then severe burning pain, even when finally ordered to become active and attack. Consequently, these soldiers were already injured before going into combat. During that winter, the US casualties admitted to hospitals because of cold injury (both trench foot and frostbite) numbered as many as 17,457 in one month, and for several weeks approached 50% of all battle casualties (Leigh 1946).

There were complaints against Colonel Doriot because of these casualties at the Battle of the Bulge until he took the leading officers to his model soldier under glass and showed them what he had recommended very early in the war. Doriot continued to advocate for proper clothing and footwear for the soldiers, and finally received support from Gen. George Marshall.

## Contribution to the Battle of the Atlantic

Our sailors and flight crews were at high risk of cold injury from prolonged immersion in fresh or salt water. When the troop ships were torpedoed, survivors might float in the water for quite a while before they were rescued. The crew in aircraft that were shot down were often left floating in the ocean for hours. They suffered from "immersion foot," which, as mentioned above, is almost identical to frostbite and trench foot. The military supplied rubber exposure suits to some personnel on ships and aircraft; these were to be pulled on over their clothes during emergencies in the water.

In 1943 we were sent a few dozen of these emergency exposure suits and asked to don them under various conditions and then give suggestions for a redesign. The studies would simulate circumstances where a ship might be about to sink, or an air crew might have a forced landing in dangerous waters; such crews could pull the suits on over regular clothing. The exposure suits were designed to be closed tightly around the face and neck so that water would not penetrate and the person would be able to float on the surface. Our testing had to involve not only cold weather exposure, but also the buildup of moisture inside the exposure suits.

With a number of the soldiers and myself as subjects, I did the following experiments: we simulated men working on the deck on a mild day who must evacuate a ship in water at 60°F, then in water of 40°F, and then at various temperatures all the way down to -40°F. We followed the typical protocol of dressing in military uniforms, putting on the exposure suits, walking on a treadmill in our hot chambers and then going and sitting in the cold room at various temperatures. Of course, we experienced all types of discomfort. Such experiments had already resulted in lengthy reports indicating how long these uncomfortable situations could be tolerated and how they might affect a soldier's ability to fight against an enemy.

We measured the amount of moisture accumulated during the work and then what the insulation would be after it became saturated. Of course considerable moisture collected, even when the outside temperature was -40°F. That was exactly the type of information that the military wanted to know; specifically, how much was the loss of insulation and how fast would that occur after exercise, presumably on the deck of the ship?

We recorded the time it took us to get into the exposure suits and recorded what sizes were needed for a variety of types of individuals. The major problem of design we identified was that it was difficult to get into the rubber suits while wearing regular clothing. This assumed there would not be time to modify the garments the survivors were wearing. We did get feedback later that while many exposure suits were issued, it was only in very special circumstances that there was time to don them before evacuating into the water. We also received information that in some cases where the exposure suits needed to be used in cold regions, the troops would not have survived without them.

An essential theme in the study of the footgear and major experiments like wearing an exposure suit in very cold conditions always involves the recording of vasomotor function, which is the constriction or dilation of blood vessels. At times the blood vessels in the skin are *vasoconstricted* and at other times they are *vasodilated.* We expected to make measurements of the activity of the skin vessels in every experiment during exposure to the cold. One of the reasons was that the human body has an emergency mechanism to try to prevent it from freezing. This mechanism consists of episodes of warm blood surging into vessels that were constricted due to cold. These episodes of vasoconstriction followed by vasodilation are called *Lewis waves* after the person who discovered it, or *cold vasodilation*. In many of our reports, we had much to say about this automatic protective mechanism of the body which tries to prevent cold damage caused by decreasing blood flow to the skin.

**Folk in the Fatigue Lab getting human subject data from respirometer.**

Before and during the war, two important principles came out of the Fatigue Laboratory, namely the acclimatization to heat and the acclimatization to cold. Although initial experiments on the later were done in our cold chamber, the best experiments on cold acclimatization were done later at a laboratory at Fort Knox where more extensive facilities allowed for long-term cold acclimatization studies.

## Field Studies

An important and unique feature of the Fatigue Laboratory was its custom of checking the artificial experiments done in environmental chambers against rigorous testing in the field. The rule had been to be prepared if necessary to pack instruments and take measurements in the field. There had been a long tradition of this, such as Bruce Dill's heat studies in the Nevada desert and successful high altitude expedition in the Andes. Robert Johnson developed field kits for doing blood analysis during field tests in both the heat and the cold in Saskatchewan.

One leader in this tradition of making physiological measurements in the field was Per Scholander from Norway, a distinguished investigator who often worked at the Fatigue Laboratory during the war. He led various field excursions from the Lab, and called this area of work *Expeditionary Physiology*.

Later, carrying this inclination through to fulfillment, Scholander, together with Laurence Irving from Swarthmore College, established a physiological laboratory at Point Barrow, Alaska. Donald Griffin and I were just two of the many physiologists who would later venture to Barrow to work in this laboratory.

I was involved with quite a few field tests while I was at the Fatigue Laboratory. One of these was a study of racial differences in the heat. The army had asked the question, "Would those of African descent be better able to tolerate wearing decontamination suits in the heat?" We traveled to an Army post in Alabama to conduct heat trials where we would compare results from 100 white soldiers and 100 black soldiers walking a mile in very hot sun while wearing the impermeable decontamination suits.

First, we administered physical proficiency tests using the Step Test, which had been developed at the Fatigue Laboratory. For this part of the study, we found that many black subjects received a higher score in physical proficiency than white subjects. Another test was to complete a one-mile walk in very hot sun, and more white subjects finished that test than black subjects. These initial measurements were followed by the key test, which was walking in the sun in an impermeable decontamination suit. In that test, there was no difference between whites and blacks.

## *Canadian Cold Weather Work*

An important field test I was involved in was connected with the Canadian Army's winter field trials held in Newfoundland in 1944. I did not have a passport and there wasn't time for the government to issue one, so I carried a packet of letters with three signatures: Col. Georges Doriot, the Canadian Minister, and the British Consul-General. Fastened together with sealing wax, golden government seals, and tied with a red cord, this impressive bundle granted passage across borders not only for me, but also for the "secret equipment" I carried. Although it sounds rather cloak and dagger, that equipment consisted of prototypes of gear and devices developed at the Fatigue Laboratory, such as a respiratory air warmer for use at -40°F.

The Quartermaster Corps issued me a military uniform with its own specially designed arm patch that read "Scientific Consultant." This odd insignia mystified some of the Military Police at border crossings, but while they were openly curious, I was never challenged or detained.

I was to be in Newfoundland for several months to work with an international team to evaluate equipment to be used by soldiers camped in deep snow with no shelter. Made up of men from Canada, North America and Britain, it was a very stimulating group to be with. I lived in barracks with the officers in a hilly evergreen forest at the base of a mountain. A Canadian infantry Colonel had brought approximately 100 troops to be our subjects. We were to teach these soldiers how to camp and survive wintertime mountain warfare.

During the week we went on maneuvers with the troops. Sometimes these were long trips where we tested equipment during bivouac, or temporary camps where the men slept on the ground using improvised shelters.

One of my most exciting assignments was to accompany flight crews in the Royal Canadian Air Force's famous "flying boats" (Canso aircraft) as they flew patrols along the coast looking for German submarines. I always had to wear the uniform in case our flying boat went down and we were captured by the enemy. My function while making these flights was to record the meteorological environment within the cabin and to recommend improvements in the clothing of the flight crew.

A member of our team in Newfoundland was Henry C. Bazett. An Oxford scholar who had served as a British medical officer in World War I, he was also a professor at the University of Pennsylvania and one of the team who had earlier developed the system of *clo* units to measure insulation. But above all, Bazett was a physiologist who liked to be in the outdoor environment.

Another companion in this endeavor was polar explorer Paul Siple, who, as a Boy Scout had gone to the Antarctic with Admiral Bird. The two of us went into the field with troops to test and redesign sleeping bags and pup tents. Our knowledge of the serious problem with trench foot during the Battle of the Bulge made us both especially interested in improving foot gear for our troops.

**In uniform in Newfoundland.** ***L to R:*** **Alan Traloar, Ed Folk, Charles Snow.**

Paul and I were wearing what we had worn on many expeditions—the shoepac built by the thousands by LL Bean in Shreveport, Maine. We knew from earlier trips that if the soldiers had to keep on their standard-issue boots for several days at a time, the wool socks would become laden with moisture. During some of our marches, both Paul and I put an impermeable barrier next to the skin in order to keep the insulation dry. After trying this we asked some of the troops to use the same procedure.

I remember seeing Henry and Paul together examining the feet of our troops for the effects of wearing an impermeable barrier while marching and camping in the cold. We did not find any damage to feet treated in this fashion, even over long marches. This information went into an in-house Fatigue Laboratory report (Siple, Bazett and Pring 1944).

In late February in Newfoundland, we witnessed a spectacular sight when icebergs and pack ice were blown inshore at St. John's, blocking the harbor entrance and shutting down the port for several days. Of course this was more than frustrating for authorities and shippers; the ice was so massive they couldn't tow or push it away, but had to wait until it was dislodged by ocean currents and carried away.

Because I was there until mid-April, I was able to see the first spring birds arriving to build nests and lay their eggs. I saw eider ducks and harlequins, lapwings, redpolls, pipits, and ivory gulls, a white-winged dove, and that harbinger of spring, the robin.

### *Scotsman's Dagger*

My roommate in the barracks was Josh MacFarlane, who was a Captain in Britain's Royal Regiment of Scotland. He had been assigned a jeep, and when we were not camped with the troops, we used it mostly on Sundays to go to the service at the Anglican Cathedral in the city of St. Johns. It was about a thirty minute drive.

The first time we went there Josh was in full dress uniform—it was also the first time I saw him in kilts. We were late to church, and as we were led down the aisle by an usher to the only remaining seats in the front row, I was puzzled because all the eyes on the left side of the church were not on Josh's face or on his handsome kilt, but on his left leg. Later, I saw that an essential part of his Highland dress uniform was a dagger worn on his left leg just below the knee; its blade and scabbard were pressed hard against the skin, almost hidden underneath his stocking with only its ebony hilt visible at the top.

Afterwards I learned that this little knife is called the *sgian dubh* (pronounced "skene dhu"). It was typically worn on the right leg, but since he was left-handed, Josh wore it on the left. In Gaelic, *sgian dubh* means "black knife," which refers to the knife's ebony or bog oak handle…as well as to its history of "dark use." It is a partially concealed weapon, after all, and whether the stares were of admiration, curiosity or dismay at seeing a weapon in church, I'll never know.

After three months in Newfoundland, often working in deep snow, high winds and at altitude, we all returned to our home bases.

Sometime after we returned, Josh visited Cambridge and stayed at my home. I had told my two children, Victoria and Chris, about his *sgian dubh,* and so when they met him, there was disappointment on their faces because there was no dagger on his leg. Indeed, he was wearing his everyday uniform, not his Scotch Highland Dress uniform. The next day he was to tour the Fatigue Laboratory, and as we left after breakfast, my children were gratified because Josh was wearing not only his kilt, *but also the dagger*—the outline of its sheathed blade conspicuous under his stocking, ebony handle exposed over bare skin just below his left knee.

## *Pemmican on Pasque Island*

Another field study related to the war effort was to test a specialized emergency diet. Our laboratory was associated with Vilhjalmur Stefansson, the great Arctic explorer who had spent five years living with the Inuit. (It is said that because he measured the heads of all the people in that region he was known by them as "Head Measure"). Stefansson described the Eskimo diet as being 50% fat and 50% protein, and during his time in the Arctic, he had discovered the curative effects of fat, which later came to be known as the need for *the essential fatty acids*. Because of his experiences with the Inuit diet, he urged the military to use pemmican—a concentrated mixture of dried meat and rendered fat—as an emergency ration for the troops.

Of course the Fatigue Laboratory was selected to test the idea. This meant that the staff and the soldiers were to have a distinct break from routine laboratory work to go into the field and live for seven days on pemmican while blood samples were taken to study the effect of this extreme diet. Because of what you might call the distractions of civilization, the study was to be done on a remote island near Woods Hole, Massachusetts.

The first island in the chain that forms the area referred to as Woods Hole is Naushon Island, and second is Pasque Island. The two are separated by a strait known as Robinson's Hole. This had been a favorite fishing spot of an elite club that met there to catch striped bass. Since that species of fish had disappeared from the area, the elaborate buildings of the club on Pasque were abandoned and it was an ideal locale for several soldiers and staff members to do the experiment. We were transported by a lobsterman named Olsen, who lived on Pasque

Island and was considered to be the lobsterman who lived the farthest from the mainland in what was said to be the oldest house in New England.

The experiment began, although after the second day, all of the soldiers refused to eat the pemmican, saying they preferred to "go hungry rather than eat the stuff." It made some of them nauseated and left a waxy layer on the roof of the mouth. The Fatigue Laboratory staff did manage to eat the pemmican for the entire 7-day study. All subjects developed severe ketosis from the high-fat diet, or from not eating at all. Thus, it will come as no surprise that the resulting report (Consolazio and Forbes 1946) did not recommend pemmican as a military ration; but at least we had paid homage to Stefansson.

Initially, I was one of the subjects of this experiment, and two days into it, could still tolerate the pemmican. But then Olsen showed up with a message that my mother was gravely ill and in the hospital in Boston. At the same time, another subject and staff member, Erling Asmussen, was notified of a family crisis, and we both withdrew from the experiment. We left Pasque Island together, rowing across the strait to Naushon Island, where we then walked several miles through the island's forest to the ferry dock. Along the way I learned the awful news that Erling's family in Denmark had been imprisoned by the Nazis.

Back in Boston, my mother's health improved, and after several days I was able to return to Pasque, where I did what I could to help wrap up the experiment. Although Erling and I later worked together in the Arctic and remained close friends until his death in 1991, he did not speak about his family in Denmark again, and I fear they did not survive the war.

## *Naushon Island*

As an aftermath of the experiment on Pasque Island, later that summer I was offered a cottage on Naushon Island for a week's vacation with my family. The cottage and the island belonged to the Forbes family, and from the red cottage we could see across Robinson's Hole to Pasque Island. Will Forbes, the Fatigue Lab's director, had made the arrangements and even provided us with a horse to ride around the island.

This vacation was memorable. Mr. Olsen, the lobsterman I had met during the pemmican trial on nearby Pasque Island, ferried us across the water. Olsen stopped by almost daily and provided us with plenty of lobsters from his traps. In fact, in one two-day period we ate lobster five meals in a row. We found four ways of cooking lobster: roasting, steaming, boiling and cold in salad! There was a nearby pond where Kit and I could fish for bass, and just off of our beach the fishing was good for the large local fish called Tautog.

We took turns riding double on the horse, covering the trails on this seven-mile-long island. Victoria devised a different way to travel—standing in my knapsack holding tightly to my forehead! It was delightful to be with the children away from work and other distractions.

We rode or walked through the woods under a canopy of branches from some of the largest and oldest oak and beech trees in New England. Years later I walked in Sherwood Forest in England and found the trees there to be similar to those on Naushon. Deer were abundant in both forests, and often could be seen nearby under the forest canopy. Another delight in the forest on Naushon was the very large holly trees considered to be the largest in New England.

## Animal Physiology

Although humans were the subjects of most Fatigue Laboratory studies, I found when I looked at past publications from the Fatigue Laboratory that there were reprints concerned with comparative physiology. The system seemed to work this way: although the goal was always to obtain knowledge on human physiology, the techniques were also used on some convenient vertebrates to record their physiology. For example, someone had brought two crocodiles into the laboratory and the blood was analyzed with the same technique as ordinarily applied to human subjects. Thus, several bits of comparative physiology came into our work at the Fatigue Laboratory.

It was also no surprise that animals in the free environment were sometimes studied; in Bruce Dill's classic volume, "Life, Heat, and Altitude," he not only described human physiology, but also included information on domestic animals at high altitude garnered from his expedition to the Andes. He even included a section about ants at sea level walking at different rates of speed in various temperatures.

During one of Bruce's visits from Bethesda during the war, I was sitting down and making measurements on soldier-subjects exposed to -40°F. Bruce burst in and asked,

"Do skunks hibernate?"

I happened to have looked through literature to find that information and I told him somewhat formally what I'd found:

"It is not known."

Bruce looked at me and firmly said, "Well, get going then!"

I must say that after three years of working on human subjects, but with a previous interest in comparative physiology, this gave me the confidence to plan a career in which one could do experiments on "mice or men." I eventually did the experiment on both skunks and raccoons, and found that while both stay quiet and rest in their burrows in mid-winter, neither animal hibernates.

I enjoyed doing this kind of research, and later on had a different kind of opportunity to make numerous physiological measurements on twenty-one mammalian species. To me it was the attack upon a research question that was important, and the kind of tissue being studied did not matter as long as it had an appropriate place in the hypothesis of the research.

In my second year at the Fatigue Laboratory, Donald Griffin left because he had been appointed a Harvard Junior Fellow. He continued his research from the laboratories of the Biology Department. One evening during the war he asked me for some assistance because four hands were needed. He had received a military contract resulting from a rather bizarre suggestion someone had sent to military headquarters. The person knew that many people in Japan lived in straw-roofed houses. The theory was to have airplanes fly over Japanese towns and cities and release bats carrying an incendiary object. Then the bats would seek out their usual habitat, crawling into the straw of the houses where the small incendiary device would set the house on fire.

The simple question asked of Don was: how much weight could the common bat, *Myotis,* carry as it floated downward from the airplane? Although we did determine how much weight this small bat could carry, the incendiary plan was never put into effect, although it was tried out in a model Japanese village in a military camp near Washington, DC. Years later, I was giving a lecture at a medical school and mentioned the bat story to the students. A member of the audience rather excitedly announced that while in uniform he had helped to build the model village for the test.

Another extensive piece of comparative physiology we did resulted from an elusive pigeon. While Bruce Dill was still at Bethesda he asked us to determine what laboratory animals would be useful in extremely cold weather for testing dangerous chemicals. He brought back Steve Horvath from his job at the Army Medical Research Lab in Fort Knox to work with us on this project. Because the tolerance of ordinary laboratory animals could not be found in any literature, we took the core temperatures of a series of animals exposed to extreme cold and removed them from the cold when their body temperature started to drop.

One of the pigeons we were studying escaped in our cold room while the room was operating at -40°F. At first the bird was not even missed, and when we started looking for it we could not find it in the many bits of equipment in the cold room. Three days later we found the bird, put it a cage, and studied it for days. The bird appeared totally healthy in all respects and there was no sign of frostbite anywhere on its body from the exposure to minus 40. We compiled these results with those from other laboratory animals and published the study in the journal, *Science* (Horvath et al. 1948).

## Notable Visitors

The team of workers in the laboratory was amplified continuously as the war progressed. Probably the most famous was F.J.W. (Jack) Roughton, FRS, who had been a Foreign Fellow in the Fatigue Laboratory prior to WW II, and returned as an official consultant during the war. A decade earlier, with N.U. Meldrum, he had described how *carbonic anhydrase* plays a critical role in the transport of $CO_2$ from the tissues to the lungs. Roughton had come from England as a Visiting Research Fellow in War Science and Medicine at Harvard, and he used the Fatigue Laboratory's elaborate facilities for analysis of blood to work on military-related problems such as CO poisoning and high altitude physiology. My association with him was mostly to help him to pack his equipment each time he returned to England.

There were many interesting visitors to the laboratory. One, Sir Stanton Hicks, a Colonel in the Australian Imperial Force was a scientist and MD who had advocated using scientific principles to develop military rations that were based on nutritional—not monetary—value. He visited the Fatigue Laboratory to discuss and promote his findings.

Researcher Edward Adolph studied the effects of thirst, and came often to the Fatigue Laboratory to exchange ideas. He had spent several months at the Fatigue Lab a few years earlier, and modeled his own laboratory at the University of Rochester based on those experiences. He later became President of the APS.

One day when I was alone in the laboratory (the rest of the staff were in the field), a distinguished-looking man with a Van Dyke beard walked in and said he would like to look around. This was Sir Hubert Wilkins, the famous Australian-born war photographer and polar explorer; he had been knighted for making the first trans-Arctic flight (in 1928) from Barrow, Alaska to Spitsbergen, Norway. He also commanded the first submarine to pass underneath the Arctic pack ice, and navigated the first airplane to fly over the Antarctic continent on a mapping and photography mission. (Much later, in 1970, I visited the Antarctic island where his plane had taken off; the volcanic lava was still marked with the lines of white lime that delineated the outline of his runway.)

At any rate, I showed Sir Hubert through the laboratory and he witnessed some of our soldiers in a cold chamber set to -40°F testing the insulation in a new type of winter uniform. I learned that he had been hired as a consultant to the Army Quartermaster Corps on hot and cold weather clothing and survival techniques; he also advised other military departments on aviation and submersible craft research (Byrd Polar Research Center Archival Program 2010).

This exchange of talent worked both ways because the Fatigue Lab also loaned its staff to the military. Harwood Belding, who had built the first thermal mannequin for the US military, was given the rank of Major and sent to England in an advisory capacity. He was in London during the Battle of Britain, and described his response upon hearing the V-1 "buzz bombs" overhead. These small, unmanned, airplane-like bombs—the first cruise missiles—were programmed by the enemy to target specific locations in London; they made a putting, or buzzing sound until they neared the target, at which point they became silent as they homed-in on the target. Alerted by the sudden silence, Woody's procedure was to scramble under a bed, knowing a bomb would soon explode nearby.

But the V-1 flying bombs were soon followed by something even worse, the V-2 rockets. These were more sinister because they were silent and exploded without warning, almost causing London's morale to fail. These ballistic missiles almost won the war for Germany; the tough Brits had tolerated bombs dropped from planes and the buzz bombs, but they could not tolerate these silent killers—the war ended just in time.

## Endings...

Finally, the war ended in 1945, and people went back to their own lives. One person who returned from the military was my boyhood friend John Usher Monro. John had been awarded a Bronze Star for his service during the war.

John Monro was persuaded to return to Harvard to assist other veterans who wanted to go to college under the GI Bill. He eventually became Dean of Liberal Arts at Harvard, and when I called on him in his office, we looked out at the statue of John Harvard. Later, in the 1960s I often thought of John Usher Monro when students rioted in Harvard Square because it was his job as Dean to show up at these riots. However, since John had seen action as a Naval Officer in the South Pacific, he was probably ready for a student riot. Ultimately, he left all that behind to devote the rest of his life to helping rebuild disadvantaged black colleges in the South, where he happily taught for years.

One day an express delivery van drove up to our house. A large crate with holes for air was unloaded. "Hans is home!" I cried to the family. Hearing my voice, that boxer dog almost tore apart the crate trying to get out. When free, he stood with his front legs around my neck, licking my face.

Hans had been "sent to war" because he was not good with children (other than our own), but we hoped he would have outgrown that tendency and could readapt to civilian life. An unfortunate incident soon occurred, however, that showed us that it wouldn't work. One day when the newspaper was tossed into the yard, Hans grabbed it in his mouth; a neighbor boy—who barely knew the dog—decided he would retrieve the paper to keep it clean for me to read later. As the boy reached down to take the rolled up newspaper from Hans, the dog bit him.

It was not a vicious act, but of course I couldn't let anyone else get bitten so I contacted the Humane Society for their advice. They said they would place him on a farm, and promised to keep in touch with us about his whereabouts. Very sadly we parted again, but were relieved when they called to tell us he had found a home aboard a freighter, where he got on well with the ship's crew; it was a good match.

We had a house and yard on the outskirts of Cambridge, and it wasn't long before another dog came our way. My sister Barbara asked us to take Trinka, their young female Boxer, because they were moving. Barbara's husband, Harold Howe, was taking a new job in New York City, which meant they had to move from their place in the country to a city apartment where they could not take their dog. We were happy with Trinka, a sweet dog who liked to sneak up into my lap.

The Fatigue Laboratory was usually too busy to accommodate a dog companion, but when I had to go to the lab on weekends I sometimes took Trinka with me. I rode my bike along the bank of the Charles River and the dog ran along beside, crossing the Lars Anderson Bridge that led to the Business School campus. Usually though, I left Trinka at home, with instructions:

"Guard the grave of Mary Baker Eddy!" I would joke. This was because Eddy, the founder of the Christian Science Church, was buried next door—in Mount Auburn Cemetery, which ran alongside our property. Eddy's mausoleum was rumored to have a telephone inside "so her followers could call her," and for four years I wondered, "What would I do if I ever heard it ringing"?! Years later I learned that a phone had been left there accidently by construction workers.

Actually it was ironic and even sad that we lived near this monument to the founder of Christian Science. My wife Dorothy had suffered greatly as a child because her mother and her aunt were both Christian Scientists. Dorothy told horror stories about her childhood, about being so ill with mumps that she was almost delirious with pain, and her mother simply calling in a Christian Science Practitioner, someone who would pray solemnly, lean over the bed and say to the sick child, "Just hold the right thought and you will be well." Dorothy's father was a well-known Boston dentist whose patients included J.P. Morgan. Obviously, her father did not share his wife's philosophy. When her parents divorced Dorothy was unfortunately left in her mother's care.

When I later met Dorothy's mother and aunt, I didn't like them much—they both exuded disapproval about everything and everyone. I liked them even less when I realized the pain they had inflicted on Dorothy; it is one thing when an adult chooses to suffer instead of seeking medical help, but to make a child suffer is abuse.

The Mount Auburn Cemetery, despite its infamous resident with the telephone, was a vast and wild place that was perfect for our son Christopher. Chris, who we also called "Kit," needed such an environment. Kit gravitated to nature, the wilder the better. Day or night it drew him and he never let a closed gate stop him; he'd simply scale the cemetery's tall fence, roam its 175 acres, and catch turtles in the big ponds.

Trinka hadn't been with us long before we heard from Barbara that Harold was retiring from his fundraising job after exceeding all expectations. It seems he had pulled off a major coup when he'd invited none other than Winston Churchill to speak at the fundraising kickoff for a major northeastern university. To everyone's surprise, Churchill had agreed, taking the opportunity to tour the US.

It was during that trip that Churchill made the famous speech at a small southern college where he accused Soviet Russia of drawing "an iron curtain" across the continent of Europe. Thus the phrase, "behind the Iron Curtain," became a rallying cry against tyranny throughout the Cold War, the period from the end of WWII until 1989.

Harold told me that his university's goal had been to raise $20 million and officials announced that they had attained $22M. However, Harold later confided that they had really garnered $24M but did not want to announce that figure because they were about to begin another fundraising campaign. Although the university shall remain nameless, I will tell you, it was NOT Harvard!

When Harold and Barbara said they were moving to a small town in the Midwest, I knew it did not bode well for us vis-à-vis our dog. Of course, they wanted to take Trinka back with them.

"No, we want to keep her," I said firmly.

But my sister came in person to plead her case. And Barbara was so determined, that when she went home, Trinka was by her side.

## Beginnings...Back to Grad School

I wanted to work towards my PhD at Harvard while continuing on the staff of the Fatigue Laboratory, but I had to overcome two hurdles: the first was to be allowed to work half time at the Laboratory so I could have an income to support my family. That was approved, and thus I could follow the precedent set by Sid Robinson and Steve Horvath, who had both received their PhDs while on the staff of the Fatigue Laboratory. (Both went on to become department heads, Robinson at the University of Indiana, and Horvath at the University of California, Santa Barbara.)

The second and larger hurdle was to be accepted into Harvard's PhD program. Applications were up because there was a constant stream of people returning from the military trying to get into graduate school. We heard that Harvard was accepting only one out of ten candidates.

I had already decided that after finishing graduate work I wanted to put 50% of my time into teaching and 50% into research—my several years of teaching bright young students at Salisbury and St. Mark's had convinced me that I would enjoy a life of teaching. That was fortunate because, unbeknownst to me, graduate school administrators realized they were facing a shortage of teachers because many students had learned about the respectable salaries paid in research laboratories. Naturally, universities did not want to train PhDs who planned to leave academia for higher-paying jobs.

I was interviewed by a committee in the Department of Biology and asked many questions. To my delight, I was admitted into the program. Later, Jack Welsh told me that had I not stated my intention to teach half time, I would not have been admitted. I never asked him to elaborate further!

Under some circumstances, the work for a PhD might have been at the Department of Physiology at the Harvard College of Medicine. However, it was much more convenient and better for everyone's schedule to do the work in the Department of Biology on Divinity Ave in Cambridge, which was convenient to the Fatigue Laboratory.

The next major question was what topic to select for a thesis. My PhD sponsor was John H. "Jack" Welsh, who had been my advisor and mentor throughout. Jack was one of the pioneer researchers in the area of neurotransmission, and the approach in his laboratory was to use bioassay techniques. If there was a material that could be weighed out that he suspected of being a neurohumor, he could dilute that material 100,000 times and still get a response on the correct sensitive tissue, such as a strip of muscle from the common leech or the nerves of the heart of the clam.

For this work on the nerve impulse, Jack needed individuals with a copious background in chemistry, which I did not have; I was not a good chemist. Instead, he encouraged me to take a problem from his earlier work on biological clocks as a thesis topic. I was delighted with that suggestion because I had already done some work in this field when I studied the effect of day length on bat reproduction. Thus, in the laboratory I began to study the effect of constant daylight on the biological clock. I enjoyed working in that field so much that I have continued doing experiments on biological clocks throughout much of my career.

My PhD thesis was the analysis of the physiology and behavior of white rats raised and maintained in continuous light. When the time came for me to publish my results, I automatically placed J.H. Welsh as the second author. But when Jack was reviewing my manuscript, he sent me a note saying, "Please take my name off, I have always considered that the thesis is the graduate student's own property and have never allowed my name on those publications."

Need I say again what a great mentor Jack was?

## Research Continues as Fatigue Lab Winds Down

Physiological experiments continued at the Fatigue Laboratory supported by military contracts (and later by a new research organization called the National Science Foundation). One interesting experiment we did for the military while I was working on my PhD was to develop an emergency lifeboat ration. The Fatigue Lab had done numerous studies on nutrition over the years, and finally we were asked to select and test a ration to be put in the lifeboats of ships. It was a hot summer and seven of us immediately went on the prescribed diet, which included a small supply of drinking water.

Although body size varied among the subjects, we did not assign the water on a per weight basis, but apportioned it equally, because we believed this is what would happen in a lifeboat. Two of the subjects were large, athletic men who happened to be medical doctors—one was Robert Darling. I was very interested in any changes in attitude and outlook as the seven days went on, particularly since we had so little water. I would say there was a graded response, with the smaller subjects affected the least and the larger individuals suffering the most. That in itself was not surprising, as they had consumed less water per pound of body weight. I was one of the shorter individuals and did not feel particularly bad, but I was shocked to watch Robert, one of our most brilliant scientists, lapse into slurred speech and a very unsteady walk. Although no overt mental decline was obvious, I would expect critical tasks such as decision-making would also be impaired.

During this time, I had several opportunities to work with Bruce Dill, who was back from Washington. He started a vigorous experimental program; his subjects included me and about eight paid subjects who were students Bruce had recruited. We were to walk fast in the hot sun for an hour or longer, after which Bruce would take a blood sample. It was summer in Cambridge, and the day was a scorcher! As the only staff member, I led the subjects at the rate requested, and came in at the head of the column—some students weren't used to such exercise in the heat. Then we all lined up to have what I think was an arterial blood sample extracted.

I should mention here that Bruce wasn't asking us to do anything he had not done. Throughout his long and productive career, Dill often served as his own first experimental subject, sometimes exercising to the point of fatigue and measuring his own responses to exercise, heat, altitude, and finally, aging. He made eleven expeditions to altitude, his last at age 87.

Bob Kark and I next served as Bruce's subjects testing boots on the Fatigue Lab's treadmill. This was the same treadmill upon which studies were done on the ability of a famous marathoner, Clarence DeMar. He had won the Boston Marathon seven times (still a record in 2011), and the Fatigue Laboratory had analyzed his heart rate, stroke volume and sweat rate at various running speeds. Our boot-testing runs were much more humble.

In both his professional and personal life, Bruce Dill could be a hard driver—but he also had a softer side. I remember one weekend at the Dill's old farmhouse in New Hampshire, where Bruce and his wife, Cloris, enjoyed having staff and students visit. My family had been invited to spend the weekend. Bruce never seemed to relax, so he and I spent the waking hours chopping brush in the woods behind the farmhouse. But, during rest stops, Bruce took every opportunity to scoop up my daughter Victoria and bounce the delighted child on his knee.

It's no wonder that many who knew Dill felt affection and loyalty for the man, along with respect and admiration for his accomplishments. I consider myself extremely fortunate to have worked with a man who is often referred to as a "scientist's scientist."

I was writing my thesis when Bob Kark designed another experiment to be run on the treadmill. Bob wanted a detail of a chemical change during running that required changes in blood glucose. The technique had been developed whereby a drop of blood could be measured for the glucose. Because any staff member could be asked to be a subject, and because there weren't many others around, I soon found myself on the treadmill day after day for the experiment.

The drop of blood was taken by lancing the end of the finger and as I walked fast or ran on the treadmill, finger after finger was lanced for the valuable drop of blood. Of course it was necessary to go on to the other hand. The next day, Bob started back on my first hand and went all the way through my ten fingers again. The same occurred on the third day. This was rather routine behavior for our laboratory, and staff often commented dryly that it was difficult to walk across the lab without someone asking you to sit down so they could make an arterial puncture for a sample of blood. The experiment with Bob was satisfactory except for one thing: each evening I was typing my PhD thesis. By the time each finger had been lanced three times it became much more difficult for my fingers to press the keys on the manual typewriter.

The days of the Fatigue Laboratory were numbered. L.J. Henderson had died in 1942, other major supporters in the upper echelons of Harvard had retired, and President Conant had never been one of the Laboratory's backers. Although government support was available, University policy prohibited such funding after the war. Thus the staff was disappearing to accept positions at other universities. W.H. Forbes went to Harvard's School of Public Health, and R.E. Johnson left to continue his nutritional research at the Army Medical Nutrition Laboratory. Fred Sargent left to work with Johnson, and later founded the interdisciplinary International Biometeorological Society, a group that flourishes today. Bob Darling joined the College of Physicians and Surgeons at Columbia University. Harwood S. Belding became Scientific Director at the new CRL in Natick, MA, where he continued to design and improve the thermal copper mannequins he had developed.

Donald Griffin was probably the staff member who became most famous after the war. I felt he was the most imaginative researcher at the Fatigue Laboratory. He was later elected to the National Academy of Science as an outstanding Animal Behaviorist, an honor richly deserved.

Bruce Dill received the Legion of Merit for his military contributions during World War II, and formally left the Fatigue Lab in late 1946 to become Scientific Director at the Army Chemical Corps' Medical Laboratory. He taught at Harvard for many years as a visiting lecturer, and served as President of the American Physiological Society in 1950. He later established the Desert Research Institute at the University of Nevada, where he worked until his retirement in 1979.

As Dill walked out of the Fatigue Laboratory for the last time, he turned to me and said, "Ed, you're in charge!" Thus casually elevated, I became the Lab's unofficial director, signing the payroll and performing various sad tasks associated with closing this iconic lab.

But although the Harvard Fatigue Laboratory ceased to be, remarkably, "its stature has become enhanced, rather than diminished, with time," observed Charles Tipton (1998). The Fatigue Lab not only contributed to the Allied victory in WWII, but also inspired the design of new laboratories and introduced a new approach to the investigation of human physiology.

**"Woody and the Copper Men." First developed in the Fatigue Lab, Harwood Belding's designs evolved throughout his career and were widely used in both business and the military. (Photo: U.S. Army, 1972.)**

## Finally, My PhD!

In June of 1947, I received my PhD from Harvard University at a very unusual commencement. Our speaker was General George C. Marshall, who was to receive an honorary degree that day. During the war, Marshall had been Army Chief of Staff under President Roosevelt, and had been recently appointed Secretary of State by President Truman. During his speech Marshall declared, "I propose that this nation assist the recovery of the two countries with whom we were at war, Germany and Japan." He even recommended specific sums of money to be put to this use. The European Recovery Plan soon became known as the "Marshall Plan," and in 1954 George Marshall was awarded the Nobel Peace Prize. It has been amusing to read again and again that the Marshall Plan was announced at "that famous Harvard Commencement of 1947."

Afterwards, I joined Woody Belding in Natick. There I was asked to review the design of the animal quarters for the new Climatic Research Laboratory. I enjoyed working with Belding and the salary was good, but I had accepted that position with the caveat that I was looking for a teaching position at a university. I could not forget the promise I had made to the PhD acceptance committee at Harvard that I would teach 50% of the time after I graduated.

Thus, while I was doing experiments for Woody Belding, I was actively searching for an academic appointment. Although I had one offer in Montana and another in Illinois, my circumstances were such that I had to ask for assistance in moving my family, and neither of those universities could provide that. After working almost three months at the CRL, I was offered a position at Bowdoin College in Brunswick, Maine, with a stipend to move my family the 150 miles northeast. Very reluctantly and even sadly, I told Woody I had to resign my position at Natick. He was supportive, so much so that he gave me a contract to continue the work I had begun at the CRL; we agreed that I would commute from the college in Brunswick one day per week.

# *Chapter Seven*

## *Bowdoin College and Kent Island Years*
**1947–1953**

### A Faculty Position in Maine

IN THE FALL of 1947, I began a new adventure. I had by now been an undergraduate student, a teacher at Salisbury School and then at St. Mark's, a graduate student, a research scientist at the Fatigue Laboratory, and finally a PhD student. Now, with my recent degree, I would return to teaching and the design of a new laboratory. Along with the duties of my first faculty position at Bowdoin College in Maine, I would have another task, which was to manage a scientific station on an island in the middle of Canada's Bay of Fundy. With that job, one could say that my focus shifted from bats and mountains to sea birds and oceans.

During my initial interview, Alfred Gross, a senior professor and well-known ornithologist, described Bowdoin's program on Kent Island. Alfred and his son, Bill, had helped develop the Scientific Station on the island. There had been an old farmhouse, but no inhabitants except for a few sheep. They had arranged for construction of a few more buildings for dormitories and educational programs. There was also a meteorological station on the island. I was surprised to learn that Fred Sergeant, my good friend from the Fatigue Laboratory, had helped establish it and often visited the island.

Raymond Paynter was the other "key player" on Kent Island. Also an ornithologist, he was a Bowdoin alumnus who was a graduate student at Yale. He was currently managing the biological station and spent summers on the island studying the birds. Ray had one remaining summer at Kent Island before leaving for his new position at Harvard, where he would become a Curator of Ornithology in the Agassi Museum of Natural History. Prof. Gross told me that I would take over the management of the station when Raymond left.

While I was deciding whether to take the position, Alfred transported me to his summer cottage a short distance from the college. I remember distinctly several events during that interview. He had a beautiful porch that looked out upon Casco Bay. There were some rather old lawn chairs on the porch, and I was assigned to one; as soon as I sat down, it collapsed to the floor.

After I recovered from this accident, Alfred asked me to see some of Casco Bay in his outboard motor boat. He wanted to show me some nesting gulls on one of the nearby islands. We had circled the island and started back to shore when the motor failed. I don't know how long we were drifting there—being stranded always seems longer than it actually is—but after many attempts to start the motor, it finally started up again and we returned to land.

In spite of all these setbacks, I said I was still interested in the position, and was sent to interview with Bowdoin's president, Casey Sills, who was a much beloved leader among presidents of small colleges in the United States.

The salary was much less than I had been making at my last position. During my meeting with Dr. Sills, I was quite surprised by an alternative he offered to me. The position I had applied for carried the title "Assistant Professor," and offered a small, nine-month salary. Sills went on to tell me, however, that if I would accept the same job as "Instructor," I would receive a higher salary. I could only surmise that the reason for this nearly unbelievable offer was that, if I accepted the lower-ranked position as Instructor, he could hire another Assistant Professor in a different field. This juggling of positions in a college may not be too unusual, but it seemed peculiar to discuss it openly with the applicant! Of course I accepted the assistant professorship, notwithstanding the financial penalty—I knew my career path lay in academia, and holding that title was a necessary step to eventually becoming a full Professor. Despite that odd offer, I found Sills to be a delightful neighbor and friend.

### *Family Troubles*

Brunswick, Maine was a college town perfect for people who enjoyed being outdoors. Just six miles to the east lay the Atlantic Coast with its beaches and bays, while a two-to-three hour drive west would get one into New Hampshire's White Mountains. Indeed, the back porch of our house on Hawthorne Street had a good view of Mt. Washington in the distance. For city-lovers, Portland, ME is about a half-hour south by car, with Boston, MA another two hours farther on.

We had only been in Brunswick about a month, when my wife Dorothy announced that she couldn't stand it there, and was going back to Boston. She went, but left Christopher, age eleven, and five-year-old Victoria behind. So there I was, beginning a new job in a new place, and trying to raise two children on my own. But for better or worse, ours wasn't a clean break. Over the six years at Bowdoin, Dorothy sometimes returned and tried to reprise her role as mother and wife; we would all be hopeful for a while, but she never stayed long. Meanwhile, I was teaching two courses.

I was totally unprepared to be a single parent—although, who is ever prepared for that? I had been raised with the expectation that, as a man, I would devote most of my attention to my career, while a helpmate managed things on the home front. I suppose that, like the children, I too was bewildered.

**Kit and Victoria, *circa* 1947.**

Fortunately, there were many warm and helpful ladies in town who were only too willing to become "nannies" to the abandoned children. Two of these ladies in particular practically adopted the two kids (and me). They were especially nice, and insisted on having a meal for three ready in the evenings for me to warm up when I got home.

People did their parts to keep me busy. The wife of one of our faculty was the director of the Bowdoin Community Theatre, and she drafted me to play one of the male leads in the J.B. Priestly play, *Dangerous Corner*. Although we gave only two performances, it was a good way to forget my troubles, and from time to time, even without a wife, I had a good time.

Some of my colleagues also did their best to keep me occupied following the inevitable shocks of Dorothy's departures: for example, the head of the Biology Department was a natural athlete who was addicted to skiing. Even when he had a pile of papers six inches high on his desk he would look at the snow-covered peaks in the distance and say, "Let's go!" I would rush home, get Chris and our skis, and off we would go up to Mt. Chocorua and Mt. Washington.

Of course, it was my job itself that did more than anything to keep me busy at Bowdoin.

### *Teaching Research: Of Trout, Snakes, and Mice*

I enjoyed teaching my two courses—Comparative Anatomy and Physiology—and worked hard to give improved courses each year of the six I was at Bowdoin College. But there was another aspect of teaching I enjoyed more than any other part of the instructional experience.

Seniors in Biology Honors were required to do an experimental semester during which we were to guide them through a research project. Each year, I provided a long list of possible experimental topics, and each student could select his own specialty from the list. At that time, the biology program at Kent Island was not well developed enough to have a student do a thesis there. But because I had implemented a well-equipped physiology laboratory, most of my honors students, who were all going to medical school, took on experiments on laboratory animals for their senior theses.

Many of the senior research projects that I designed were exciting, but three in particular stand out. Two resulted in published abstracts and could easily have been material for technical publication.

The first was an attempt to modify the artificial ionic solutions used to keep excised mammalian organs alive to be studied. For example, the correct Ringer's Solution for tissue taken from frogs had been well known for many years (and still is). Scientists can follow a procedure to make this solution in their own laboratories. In the student physiology laboratory, we worked with a variety of kinds of tissue from lower vertebrates, and we wanted to do an experiment using fish hearts. There was nothing in the literature to describe the correct physiological solution to use in the experiments for fish, so I set up a senior experiment to see whether modifications needed to be made in frog Ringer's to maintain the fish heart. We set out to record the beat of the fish hearts in Ringer's, and possibly in modifications to that solution.

I contacted the State Department of Fish and Game and they became very interested. They proposed to give us any number of live adult brown trout. Fish and Game farmed these fish, maintaining the adults on a diet of liver, and releasing the young into streams. As the experiment developed, a tank truck backed up to the laboratory door every day for a week, and delivering two very large brown trout in excellent condition. We prepared tanks to accept our victims and proceeded to first study each fish heart intact in the animal, recording the amplitude and rate of beat. The next experiment was to isolate the heart in Ringer's Solution, and continue to record the same parameters. My recollection is that we did a series of eight trout. We tried a modification to the Ringer's Solution and did not find a difference in the response of the heart compared with the frog Ringer's.

At the end of each experiment, we looked at the beautiful flesh of the fresh trout and decided it should be taken home for dinner. We only tried this once, however. The fish looked delicious after being filleted and grilled, but it tasted so much like the liver it had been fed, that we might as well have been grilling calf's liver from the market. From then on we decided that if we wanted liver, we would buy actual liver rather than to bother with cooking the fish, which had already processed the liver on its own.

A second project had to do with snakes. I was glad that the student selected that particular topic because of my long interest in reptiles. The project had also been inspired by the work of my old friend Oliver "Paney" Pearson, when he'd found that the saliva of the *Blarina* shrew was as toxic as that of a poisonous snake. As we all knew (thanks to John Welsh), pit vipers like Rattlesnakes have long hollow fangs, and as they bite, muscles contract on the poison gland to inject poison into the victim. Although so-called harmless snakes do not have hollow fangs, some textbooks have suggested that the saliva may be toxic. My student's project was simple: to determine whether it was or not. Accordingly, we ordered four kingsnakes (*Lampropeltis spp.*) and four water snakes (*Natrix*), all classified as "harmless." For anesthesia we placed the snakes in the refrigerator, and when they were converted to a totally anesthetic state, we beheaded them.

The student was assigned to find diagrams of the anatomy of the "harmless" snake salivary glands, and soon learned to dissect out these glands. We used the eye as a control because the eyes contain so much protein. The salivary glands were ground in a mortar with sand and saline, and the same was done with the eyes. We did four injections from each specimen, two from the salivary glands and two from the eyes. Since there were eight snakes this meant 32 injections into white laboratory mice. (Unlike Paney Pearson, we were not tempted to inject the material into ourselves!) The results were unequivocal: every mouse that received the salivary glands spun over and died within a very few minutes. Every mouse that received the eye extract exhibited no harmful effects and could be returned to the animal colony.

What followed after that snake experiment was especially frustrating to me because the results were clear cut and should have been recorded in the technical literature, if only as an abstract. However, the student went off to medical school, and I lost touch with him. I confess that, like many researchers, I was probably guilty of preferring to *conduct* experiments rather than to *write up* the results, and as a busy assistant professor, I was soon caught up in the next experiment.

The third experiment had to do with penetration through the skin. For many years I had been investigating what material can pass through the human skin into the capillaries. My student and I decided to determine whether radioactive calcium would be able to penetrate mammalian skin, using hairless mice as a model of human skin. The Bowdoin chemistry department had a license to use radioactive material, and we used the facilities at the Climatic Research Laboratory (CRL) in Natick, MA to analyze blood. We set up the experiment so that the only way the radioactive calcium could enter the mouse circulation would be through the skin, and found that it did in fact penetrate the skin, probably through hair follicles, whether functioning or vestigial.

The student who did this experiment with me was Peter Hawley, whose career was so targeted and successful that he eventually became Dean of the Medical School at the University of Illinois. During his pathway to academic success, Pete was my assistant on several significant projects, and later became my PhD student at the University of Iowa.

### *More Military Research: The Bunny Boot*

One important project that Pete Hawley assisted me with during my first two years at Bowdoin was the study I was conducting under contract with the CRL on cold weather footwear for soldiers. For this work, I commuted to Natick every Tuesday.

The armed forces wanted to improve the footgear of the soldiers required to camp in the cold, and wanted us to address a serious problem with the shoepac, which was that the socks inside these impermeable boots became soaked with perspiration and stayed that way. My task was to determine whether feet perspire more or less if an impermeable barrier is worn near the foot or farther removed from the skin. This project continued valuable work started earlier at the Fatigue Laboratory, specifically the 1944 field test in Newfoundland, where I had worked alongside Henry Bazett and Paul Siple.

I needed someone to serve as "foreman" on this important project. Soon I heard about a man who lived on a nearby island and needed a job. That man was Robert Peary, Jr., son of Admiral R.E. Peary, the famous Arctic explorer who led the first team to reach the North Pole. Bob had left a good job in Pennsylvania to move north when news reached him that his father's home on Eagle Island in Casco Bay—empty since the Admiral's death in 1920—was being damaged by vandals.

The house on Eagle Island was special to Bob not only because he had spent long summers there while growing up, but also because the island and the home his father had built on a rocky bluff, much like "a pilot house placed on the prow of a great ship" had meant so much to his father. Bob and his wife Inez were doing their best to care for the house and the Island, but without resources they were barely subsisting.

I met Bob Peary shortly after his venture of raising angora rabbits had failed when the bottom dropped out of the angora wool market. Fortunately, Bob and I hit it off immediately, and he and Inez moved to nearby Brunswick so Bob could work with me. During the project and afterward, he and his wife became such good family friends that my daughter Victoria started to call them Uncle Bob and Aunt Inez.

Bob Peary and I conducted a variety of tests under different conditions. In one field test, we asked ten faculty members from Bowdoin if they would be subjects on a trip to the mountains to test various types of foot gear. Within our shoepacs, each foot was dressed differently. On one foot we wore an impermeable barrier outside of one sock (near barrier) with two wool socks outside of that. On the other foot we wore three wool socks with the impermeable barrier outside of them all (far barrier). With eleven men dressed this way, we climbed and camped on Mt. Chocorua in New Hampshire's White Mountains. Despite it being summer, it was quite cold on top of the mountain. Simulating some of the conditions soldiers would face, we did not remove this footgear for three days. When the ten of us arrived back in Brunswick, the doctor at Bowdoin College inspected everyone's feet, which were all healthy, and he could not tell which way each had been dressed.

In our study we found that appreciably less perspiration accumulates with an impermeable barrier worn near the skin than with the barrier farther removed from the skin. Also, the amount of sweat accumulated under the far barrier was four or five times that accumulated under the near barrier (Folk and Peary 1950). Perhaps this accounts for our subjects' preference for the near barrier as being more comfortable.

Most surprising was our finding that, under the near barrier, foot perspiration was reabsorbed through the skin, thus keeping the sock and foot more dry. The implication on the future design of footwear was significant because these findings showed that, contrary to previous assumptions, evaporation is not essential inside a shoe. This experiment showed that insulated footwear that is water-impermeable on both the inside and outside of the shoe could be physiologically acceptable (Folk and Peary 1951). Our reports contributed to the development of the Thermal Boot, which had a very thick layer of foam rubber insulation.

At about that time, US troops were overseas fighting in the Korean War. They faced bitter cold and snow in the mountains of North Korea. In the fall of 1951, we faculty members who did the experiment were pleased to see the televised issuing of the Thermal Boot to US troops in

Korea. First made in black, they were quickly dubbed "Mickey" boots (for Mickey Mouse), and when a better-insulated version was issued in white (for more extreme weather and camouflage in snow), it became known as the "Bunny" boot.

By the end of the war in Korea in 1953, almost 37,000 American soldiers had died, more than 8,000 were missing in action and over 103,000 had been wounded. In my reading of Korean War history, I was gratified to find many statements to the effect, "…the only part of my anatomy which was warm were my feet because they were in the Thermal Boots."

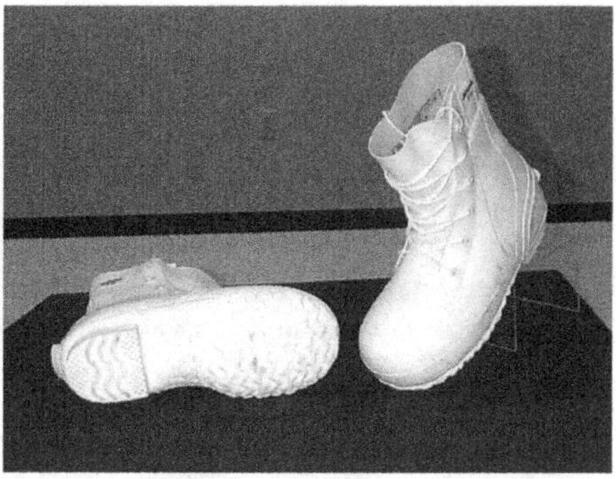

**Foam rubber Thermal Boot, later called the Bunny Boot. (Photo: US Army.)**

## Kent Island and the Bay of Fundy

Kent Island was given to Bowdoin College by amateur ornithologist J. Sterling Rockefeller as a sanctuary for nesting birds. This small, uninhabited island lies in the middle of the Bay of Fundy, a 170-mile long channel that separates the coast of northern Maine and Canada from the peninsula of Nova Scotia.

The bay is a cold body of water with a dramatic presence. Here, twice each day, the tides rise and fall by as much as 48 feet—the highest tides in the world. When the tide goes out, the ocean bottom is exposed for great distances, but when it returns, it is with such force that in some of the rivers that flow into the bay, the current is temporarily reversed, creating the phenomenon known as a *tidal bore*.

The island had a few buildings, one of which was a station for biological research. I became intimately involved with this place during my time at Bowdoin, becoming Director of the research station there.

Originally, the purpose of the program on Kent Island was to assist in the recovery of the Common Eider Duck. These sea-ducks build their nests close to the water, and keep the chicks warm by lining their nests with soft eiderdown from their bodies. Before factory farms provided domestic sources for down, eiderdown was in high demand for feather pillows and quilts. The breeding colony of this endangered species had been greatly decimated because boatloads of residents from nearby Grand Manan Island came there every spring to collect eiderdown as well as gather the duck's eggs, which were considered a delicacy, and they often took the whole nest, which further disturbed breeding.

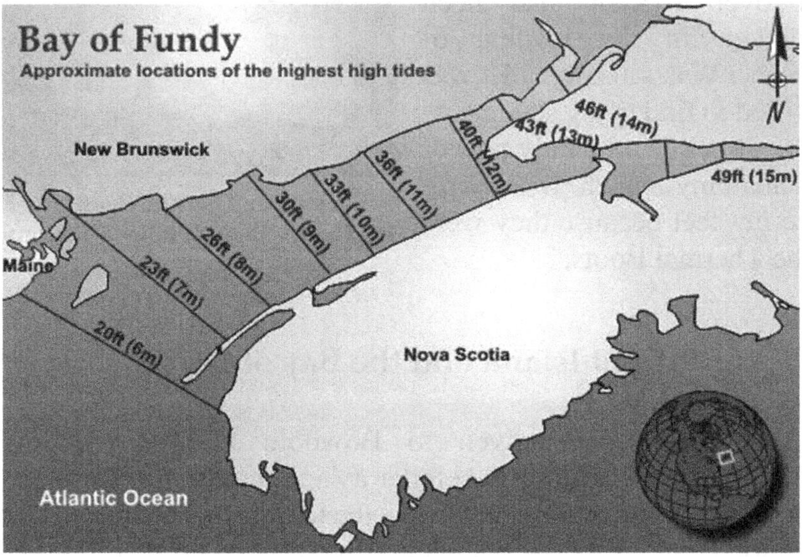


**Schematic of the Bay of Fundy and its remarkable tides. (Source: National Oceanic and Atmospheric Administration.)**

When the island was donated to Bowdoin College in 1933, there were fewer than 100 nesting pairs. The college hired a caretaker to keep the "eggers" from further impacting the duck population. Ultimately, the protections of the Canadian Migratory Bird Act were enforced, and by 1948, the eider duck population had increased to 2,000 nesting pairs.

I first visited there in the summer of '48 with three of my students and Raymond Paynter, who was still Director of the island. Kent Island is less than two miles long and a mere half-mile at its widest point, but it was an exciting and wonderful place for me as an experimental biologist. It was such a pleasure to explore the entire island with Ray. In the evenings, we banded Black Back gulls and then stopped to watch the Aurora Borealis until the fog rolled in. At night, we listened to the petrels sing their lovely warbling songs. By daylight, we inventoried the numbers of nesting Herring Gulls, one of twelve kinds of seagulls. I made plans to take core temperatures of the gulls on their nests. Ray introduced me to the station's caretaker, Ernest Joy, and they filled me in on the management of the island.

Until Kent Island, I'd had little experience with sea birds, and a particular joy to me was becoming acquainted with Black Guillemots. These birds are exceptionally beautiful in their black summer plumage—they sport a snow-white spot on the wing, have a dramatic red gape (the inside of a bird's open mouth) and vivid red legs. The bird's eggs are quite large relative to their body weight, and for years I thought their egg-to-body ratio could be a record in the ornithological world. But no—as I later learned, that record goes to the Little Spotted Kiwi of New Zealand, whose eggs, at 26% the female's body weight, are the largest in proportion to its size of any bird in the world. In any event, not holding the egg-to-body-weight ratio doesn't take away from the Black Guillemots' dramatic beauty.

### *Studying Leach's Storm Petrel*

My major new experience was to get acquainted with the petrel colony. The variety nesting on Kent Island is Leach's Storm Petrel. These birds live on the surface of the ocean, usually far away from land. They fly at night to the evergreen forest on Kent Island to dig an underground burrow and lay one egg each year; this is where they raise their young to maturity. I realized there was a lucky opportunity to study the unusual 40-day incubation of the petrel egg. When I returned to Kent Island the next summer, bringing several physiology students with me, I also brought a readout device with thermocouples, instruments we had used at the Fatigue Laboratory. The copper junction of the thermocouple could be fastened to the incubating egg, and a long pair of thin wires could come out of the burrow to a distant spot where one could read the temperature of the egg. With this device, we made many measurements of incubation temperatures.

Donald Griffin, who had studied petrels for his homing experiments, taught me how to determine when the bird was on the egg: one puts a series of fine sticks over the entrance to the burrow and then returns to see if the animal had disturbed the sticks. Fortunately the thermocouple on the egg did not prevent the bird from incubating the egg. It was possible to record the temperature of the egg before the mother arrived, then the change as she sat on it, and the cooling process after she left.

I invited Don to give a talk at Bowdoin. After his lecture, he told me privately about some experiments he had been conducting at the Naval Arctic Research Laboratory in Pt. Barrow, Alaska in which he attached radioactive capsules to the legs of petrels to study their migration. He told me he was worried that one of the local Eskimos might come across one of these birds, and unknowingly either wear the capsule as an ornament or eat the bird. These concerns caused him to abandon the study after a short time.

During the second and third summers, I was able to determine that the body temperature of the incubating petrel is lowered to a point that we can call semi-dormancy (98.96°F), and the incubating temperature taken against the egg (64°F) was lower than that of ordinary birds such as songbirds (91.6°F). In 1949 and '51, I published these findings as abstracts.

As a side note, years later in 1979, my friend Herman Rahn added an improvement on the experiment by drilling a hole in the egg and determining the temperature of the interior of the egg. What I had measured was the environment between the skin of the mother and the eggshell. Herman recorded a warmer incubating temperature within the egg (91°F) than I had recorded from its surface. The average body temperature he found in incubating adults (100.9°F) was also slightly higher than what I had recorded.

### *Managing Kent Island*

When Alfred Gross hired me, he had told me I would be the director of the Kent Island station after Ray Paynter left. He repeated it emphatically on other occasions. But, knowing Alfred as I did, I expect he mentioned this to few other people. The fact is that by his declarations, and moreover by actual practice, I did my best to manage the island for six years. It was a rewarding, but time-consuming task, and it was also quite an adventure.

Each year, I took a budget for operating the island and biological station to President Sills, negotiating items and reporting to him on annual expenditures. It was mostly transportation expenses, because the journey from Brunswick, Maine to this small island in the Bay of Fundy was a long one.

The trip took almost all day; the means of transportation and route we took varied depending on who or what was going with us, and of course it all depended on the fog, the tides, and the weather in the bay. We sometimes drove to a small Canadian sea port several hours to the north, passing near President Roosevelt's summer home on Campobello Island. From there, we boarded a ferry for the ninety minute trip to Grand Manan Island—which, the first time I saw it, looked like one great cliff rising from the sea. Kent Island lies five miles south of Grand Manan, and we were usually transported by a lobster boat captained by either Lester or Myhron Tate. Then we climbed into a small dingy to get to shore. That trip took about an hour. Sometimes one of the Tates had business on the mainland, or we specifically asked that they meet us there to take us directly to Kent Island, which made for a shorter trip. We went by sailboat once when there was an alumnus of Bowdoin who wanted to sail to the island.

**Sailboat *At Last*. (Photo: Harry Hunt.)**

The personnel on Kent Island for the first two years consisted of Ernest Joy, the caretaker of the island, and Mrs. Chase, his housekeeper. Carrie Chase had been the wife of a lobsterman who was lost at sea in a storm, and though she was officially Ernest's housekeeper, he introduced her to people with the assertion, "This is my woman."

I spent many pleasant hours talking with Ernest. He told me he had grown up fishing "by weir." This is a method in which people construct large circular traps of wood, brush or stone to catch fish; the elements of the weir are arranged so that that fish (or eels) can swim in, but not out. Weir fishing has been used around the world for thousands of years, and is still practiced in the Bay of Fundy. In the upper Bay when the tide is out, fish trapped in weirs can be picked up off the bay floor. Around Grand Manan, where the changing tides don't quite expose the sea floor, weirs are tended by boat, and nets are used to raise the trapped fish to the surface. I once had the pleasure of visiting a Canadian weir, which was designed in a spiral fashion which condensed schools of herring as they came to the trap at the end of the spiral. Many boxes of Salt Herring were sent out from that village. I noted an abundance of another fish called Pollock; they were once scorned by the local fishermen, but are now common on menus of most restaurants.

Sometimes sharks and whales entered the weirs. When this occurred, a charge of dynamite or a shotgun took care of them. Just before I arrived at Kent Island, a Right Whale had been killed and had floated out onto one of the small islands nearby. I wanted to see the skeleton, and someone took me to see it. Most of it was still upright, holding the whale's shape. The animal's vertebral discs were the size and shape of large dinner plates; I took one home and after cleaning it, used it for many years as a tray for hors d'oeuvres, to the astonished delight of my guests.

Ernest Joy had countless whaling stories. I remember one in particular, in which he had come across a sleeping whale. He had been out checking his lobster traps and had some stones with him for ballast, and he thought he would play a trick on the whale. According to his story, he rowed up to the whale and dropped a stone in its blowhole. Irritated, the whale sounded (made a deep dive), and water came into Ernest's boat without causing serious damage, and that was the end of *that* story.

Not long after I met him, Ernest told me he did not want to spend another winter on the island unless he had a working radio. He did have a radio, but its twelve-volt battery was out of charge and there was no charger on the island. For lack of any other plan, on my way home I put the battery in a knapsack on my back, planning to get it charged up at Bowdoin. As we pulled up to the dock at Grand Manan, I saw I'd made a mistake—it was low tide. Since the tides in this area vary so greatly, that meant I had about a forty foot ladder to climb. I well remember the weight of that twelve-volt battery as I climbed that ladder. At home, I charged up the battery, and also purchased a new radio for Ernest. I planned to take it or send it on the next trip to the island.

At about the same time, I had another unwieldy cargo that needed to be taken to Kent Island. Earlier that summer, I'd been to the island with another faculty member and a visitor, and we'd found that rats had gotten into the mattresses in the dormitory. Back at Bowdoin I arranged for the right size and number of mattresses, and talked to Alfred Gross about the best way to get them to the island.

By chance Alfred was to be on a ship with his friend Commander MacMillan toward the end of the summer, and they planned to stop at the island. Of course he agreed to transport the mattresses, radio, and charged battery. This plan was carried out, and the radio operator on the MacMillan ship rigged the radio and taught Ernest how to use it.

A terribly unfortunate event occurred that winter. Ernest's companion, Mrs. Chase, became ill with pneumonia, and Ernest could not make the radio work. There was a bad storm that prevented them from getting out for medical assistance. Ernest tried to attract attention by building a bonfire on the beach, but it did no good; Mrs. Chase died in the farmhouse. When the storm had subsided, despite very rough winter seas, Ernest rowed his boat to Grand Manan Island. It was considered quite a feat in view of the height of the waves. Later, Mrs. Chase's body was removed from the island, and although Ernest finished out the winter as caretaker, he found the island to be too lonely a place, and left for good the following summer.

I began looking for a person to be the caretaker on the island. I was unable to find someone right away, although Pete Hawley and I did go there for a short period to make measurements on petrels. I finally located a retired lobsterman who said that while he would not live there, he would visit the island frequently to inspect and put in appearances during the summer. He was not there during the winter, however, and on his first visit that spring, he found that duck hunters had invaded the farmhouse, used it for cleaning their ducks and left a mess. They had also entered buildings (that had been locked) and scattered the bird bands used for studying the tree swallows of the island.

The caretaker telephoned me at the college to report the vandalism, and because the island is in Canada, I felt required to report what had happened to the Royal Canadian Mounted Police. In the interview they asked if any damage had been done to United States property. I explained that the bird bands strewn about were US property, and they said this was enough for them to send an officer to inspect.

I formed a group to travel as soon as possible to Kent Island to put things in order. My father was visiting me in Brunswick at that time, so I invited him to go with us to see the island. We also took young Christopher, who I could already tell was a budding naturalist. I hired the lobster boat run by Myhron Tate, and we met at the dock on Grand Manan for the trip.

A handsome young officer in the red uniform of Canada's Mounted Police was also there to join us. The waves were choppy and high as Myhron piloted us toward Kent Island. There were five people on the boat and the four civilians did not seem bothered by the unsteady boat. However, the Mountie, who would have been quite used to the gaits of a horse, could not ride out the rough waves. We have always remembered that beautiful red uniform draped over the ship's gunnel most of the trip.

Upon arrival we inspected the damage, which fortunately was not extensive. I signed the policeman's report and made arrangements with the lobsterman to make more frequent inspections.

It soon became more difficult for me to get to the island. The reason was that the head of Biology started a summer school so the faculty members who were not independently wealthy could supplement their meager nine-month salaries. The previous two summers I had managed with the small military grant, but that was finished, and in addition to supporting a family, I also had graduate student debts. Therefore, I was forced to teach an Introductory Biology course for the entire summer; the course was always full because it helped the students to graduate early.

The next summer occasioned more difficulties with staffing at Kent Island, as a major problem came to head. During the six years I was at Bowdoin, Alfred Gross, the senior ornithologist, was semi-retired. He was seldom in the department and there was no bird course given during the six years that I was at Bowdoin. This meant that we did not have interested students to take to the island except the two years that I was able to take my physiology students there to set up experiments.

I interviewed many biology students trying to find some who would accept a salary to be assistant director and resident guardian of the island during the summer. There was only one biology student willing. His name was Al Barnes, and he had visited the island with me several times before. He, at least, was deeply interested in Ornithology.

Al had begun projects on the eider ducks, the black-backed gulls, and the petrels. However we could find no one to go with him, and he landed on the island alone hoping there would be visitors that summer. There were very few, however, and he summoned a lobster boat in August, leaving the buildings unattended again.

The next year I opened the station as usual in the spring, and closed it down at the end of the summer. At that time, the only people who were going to the island were the meteorologists who came down every summer and a few visitors.

The next summer was even busier in the Department, but I did spend some time on the island in the spring. When I returned in the fall to close the laboratory, I was accompanied by Larry Hall, the head of the English Department. Myhron Tate dropped us off at the station and agreed to come back the next day. But overnight a hurricane had come up down south and the waves were extremely high the next morning. We barely missed having to spend three or four days marooned with only a box of salt herring and a bottle of brandy for calories. Much to our surprise and delight, Myhron did return for us despite the approaching hurricane. Those were the highest waves I had ever experienced in a small boat.

## A Busy Final Year

Back in Brunswick we had been undergoing very large changes in the biology department. The science building was undergoing a major renovation, and some of the high ceilings were being lowered, to add one additional floor. We had to empty the science building, move all our teaching equipment to other buildings, and design new offices and laboratories. One of my tasks was to empty the natural history museum and store the specimens in a classroom across the campus. I continued teaching two classes, while also designing the new laboratory, and spending time at Kent Island trying to improve things there. It was a time consuming, complex year.

I did find a little time to take Kit and Victoria trekking through nature, although I wish it could have been more. On these hikes, I probably also gave in to the impulse to "lecture" on bears, eagles, birds and other wild residents. Whether I was with him or not, Kit continued to spend much of his time outdoors, roaming the woodland groves and ponds at an old Air Force base in Brunswick. He sometimes took his sister wandering through the wilds with him. During the summers, if he was not out exploring life in ponds, Kit could usually be found digging clams for Snow's Clam Chowder.

### *Visit With an Old Friend*

George Quaile, a youngster I had taught earlier at Salisbury, was now a student at Bowdoin, although he was not in any of my classes. Unfortunately his father, Emerson, had died at an early age, but George's mother, Charlotte, was living in a cottage on the ocean near Bowdoin. Of course I went to visit her, as the family had continued to be my good friends.

At Charlotte's cottage one day, I pointed to one of Casco Bay's off-shore islands and said,

"It would be lovely to go out there."

"Yes, let's go," she said, making up her mind.

I suppose my face held a questioning look because from where I stood on the beach, I thought the sea looked a bit rough. I also wondered how we would get there.

"I have a canoe!" She announced, reading my thoughts. She led me to where the canoe was kept and we carried it down to the shore, put it in the water, climbed in, and paddled through high waves to the small island. I was becoming tired, but Charlotte's energy seemed unflagging. I was amazed at her strength and stamina; she was probably twenty years older than I.

After we reached the tiny island, we looked around and rested a few minutes. Then we clambered back into the canoe and repeated the vigorous paddle back to shore. I was happy to see the indomitable Charlotte had not changed; she was still a remarkable woman!

### *Birds with Cold Feet*

One day, the renowned scientist and professor, Larry Irving, came into my office while he was on a visit to Bowdoin, his undergraduate alma mater. He was there to receive an Honorary Degree from Bowdoin College. I knew him from graduate school, where he had been an important mentor to several of my colleagues; now he was to become an adviser to me as well. He had already established an active research laboratory at Point Barrow, Alaska (where I would later also establish research quarters).

The reason he "accosted" me at Bowdoin was that he had begun investigating how birds with unfeathered appendages keep from freezing at ambient temperatures of -4°F, and he wanted me to help him study and explain how gulls, with their bare feet, could walk on ice for long periods without damage, when the air temperatures were so low.

With everything else going on at Bowdoin, I had no time to conduct such a study, but I did manage to obtain gulls that were wintering on the Maine coast and then transport them to the Harvard Medical School, where his colleagues Charles Lyman and Paul Chatfield did the experiment to explain their tolerance. Their paper (Chatfield, Lyman, and Irving 1953) has become a classic; I appreciated Larry's insistence, and was happy to have contributed.

### *A Chance Encounter, New Opportunity*

Bruce Dill had kept in touch with me over the years since I'd worked with him at the Fatigue Laboratory, and he encouraged me to go to as many national meetings as I could. It was good advice. During my last year at Bowdoin I was able to attend the American Physiological Society (APS) meeting, where I ran into Steve Horvath, Bruce Dill's son-in-law, and a scientist who had worked at the Fatigue Lab. In fact, I had been hired there to take the position Steve was vacating when he went to the Army Medical Research Laboratory at Fort Knox. I had worked with him to do some studies of comparative physiology on animals in the cold.

Steve was running down the stairs and stopped to say, "Ed, there's a job in Iowa that you should apply for." It was his position, which he was leaving for a job in Pennsylvania. Steve encouraged me by telling me about the environmental chamber he had built in his laboratory.

Thus, in the spring of 1953, I interviewed for the position in the Department of Physiology, Medical College, State University of Iowa.

Actually, I had come to the interview mostly as a favor to Steve Horvath. I did not want to leave Bowdoin, but also thought I should look at my options. I was bowled over when I was offered an associate professorship *with tenure*. Without tenure I wouldn't have considered such a move, but the offer of tenure made the difference.

The position was targeted to begin that fall. I said I would be happy to accept, but could not come until after I'd helped move the Bowdoin biology department back into its new building. The officials at Iowa agreed, and I scheduled my move to Iowa for January 1954, after the end of Bowdoin's fall semester.

Much later, when I was already at Iowa, the director of the department, H. Hines told me something surprising. He said the only reason he hired me was because I didn't want to leave Bowdoin College! While I hope he was joking about that being the only reason, he explained his hiring philosophy, saying "I never hire anyone who wants to leave where they are."

In the months before I was to leave New England, I tried to convince Dorothy to come with the family to Iowa, but she decided she could not leave the Boston area.

We got Bowdoin's biology department moved back into the new science building late that summer, just in time to begin the fall semester. I particularly enjoyed working in the new physiology student laboratory, which I had designed. It was a shame to leave it, but Steve Horvath's old lab awaited me in Iowa. The last thing I did for Bowdoin was to move a mounted adult polar bear, fixed upon a bed of artificial snow, back into the new museum.

Thus an interesting period ended in Brunswick. It had been a demanding six years, filled with responsibilities to family, students, classrooms, laboratories—and a small island. I moved to Iowa looking forward to new adventures, never expecting that there would be more and better adventures than I could have imagined!

# *Chapter Eight*

## *Joining the Faculty of a Medical School, The University of Iowa*
**1954**

### An Unexpected Start

MY NEW APPOINTMENT was in the Department of Bacteriology, Stomatology, and Physiology at what was then still called the State University of Iowa. Because I was joining the faculty in January, the middle of the school year, I had been assured I would have no teaching duties until I had time to get the family set up in a new home, and had time to organize my new laboratory.

But no sooner had I arrived in town when I received the bad news: the professor who was to teach Physiology that semester had been taken to the hospital with a kidney stone, and I would have to start teaching his class of 140 students the very next morning.

Thus, I was to take charge of a large course, lecturing four times a week and teaching in the laboratory two afternoons a week. It might have seemed quite a change from teaching small undergraduate classes at Bowdoin College to teaching much larger graduate classes. But I credit my earlier experience lecturing about snakes to big audiences around Boston with helping me feel at ease in front of such groups.

My Physiology class was an interdisciplinary course for students from the dental school, the physical therapy program, and the departments of microbiology, pharmacology, anatomy, and biochemistry. There was a separate Medical Physiology course for students in the medical school and physiology department, and I also lectured in that course. I enjoyed working with these young people and encouraging them to prepare for an academic career.

I was also working on acclimatization in the cold room of the laboratory that Steve Horvath had previously set up. It was humorous how people began to link Steve and me together; after all I had now taken two of his former positions, first at the Fatigue Laboratory, and now at Iowa.

Meanwhile, I was trying to get the family, which consisted of Vici, Kit and me, established. I enrolled the children in school: 11-year-old Victoria attended 6th grade in one of the local public schools, while Christopher at 17 was a senior in the University High School on campus. I hired a woman to be at the house after school to receive the kids.

Dorothy had not accompanied us to Iowa, but she soon arrived, making it clear to me that she had come because she had no place else to go. She did not stay in our house long, but soon moved in with her friend who ran a local bookstore. For a while, Victoria lived with them, but then moved back with Chris and me because the woman her mother shared a house with was difficult to get along with.

I had resisted getting a divorce because I did not believe one should do this when there were children in a family. However, Dorothy initiated proceedings and when the divorce came through, left the children, including her son Kit, in my care. It was difficult for all of us, and because I had such a demanding job, it was hardest on the children.

Later on, Dorothy moved to Rockport, Massachusetts. Although she tried to continue being part of her children's lives, I don't recall that she ever returned to Iowa City. She eventually died in New England.

## Teaching

I was assigned four graduate students in short order that first semester at the University of Iowa. Despite a heavy lecturing load, I felt at home with the students who asked to do PhD work in my laboratory, perhaps because of my experience with the honor student researchers at Bowdoin College. Sometimes it was a challenge, however, because to my surprise these graduate students seldom knew the topic they wanted for their thesis, or even what area of work to select. They had confidence that I would guide them to the right one. I felt a lot of responsibility for this because usually their PhD topic governed their research interests during their later academic career. The thesis topics they selected from those I had listed were in mammalian hibernation, biological clocks or a combination of the two—studying the biological clocks that still exist during the low body temperature of hibernation.

Over the years, we all received better and larger research laboratories and by the end of my career, I had trained 20 PhD students—if you include one in the Netherlands who I supervised while on a Fulbright, and helped translate his Dutch thesis into English.

My relationships with these students varied once they had graduated. In some cases, fond personal relationships had developed and we retained an almost weekly interaction for years. One former student, Abbie Moos, was traveling through Iowa from the East Coast because she had been accepted at a veterinarian school out west. She stopped at my farm for a few days with an entourage (as she expressed it) that included two horses, two dogs, two cats, and a new husband! My first PhD student, Marvin (Bud) Riedesel, became a professor in Albuquerque, NM, and later co-authored the third edition of my textbook with me. Bud and his wife Rosemary remained my good friends until his untimely death some years ago. Another former student never fails to send me Christmas cards with photos from her farm in South Carolina, where she raises and shows Miniature Donkeys, Norwegian Fjord horses, and other equines.

At the opposite extreme was the student who wrote once after he received his PhD. Later, when I wrote to that address, the letter was returned and I never heard from him again, nor could I find any record of his whereabouts.

I could go on and on about the terrific people who have been my students; I have been proud of them all.

### *Teaching Philosophy*

Years ago a significant book was written with the title: "And Gladly Teach." That title sums up my philosophy. My success at teaching boils down to a word mentioned often by many of my students: *enthusiasm*. I suppose my enthusiastic teaching style arose naturally from my genuine enjoyment of the subjects I taught.

I made my academic duties harder because I rebelled at an examination system that tested medical students on the details of physiology with multiple-guess questions. I always included a section headed, "Define and explain in two grammatically complete sentences, one of the following subjects…"

I believed in the value of essay questions—and yes, I personally read the essays or short answers of over 140 students, even when it meant staying up late for several nights to read them. Many of my colleagues gave tests that were all multiple-choice, which, when teaching a huge undergraduate course, probably makes sense. But as a student I had enjoyed answering essay questions, and as a teacher I felt it was important because it allowed for continued interaction between professor and student, which multiple-choice questions do not.

Another benefit is that short answer/essay exams show not only whether the student really understands the topic, but also whether he/she can write! It is actually unfair to college students *not* to require some grammatically correct, complete sentences. When I found students who did not write well, I always did my best to help them improve.

And there was a reward! I cannot count the times decades later when white-haired former students of mine sought me out to thank me for making them answer, define or explain essay questions.

That I truly loved to teach must have been apparent because years later (in 1983), I was named University Teacher of the Year in the College of Liberal Arts (M.L. Huit Award), and about a decade after that was given that title again, this time from the Medical College (J.P. Long Award, Basic Science, 1996).

## *A Busy, but Fruitful Period*

Despite being busy—or perhaps thanks to all this professional activity—this was a wonderful period. For ten years I did "bench work" in my lab, usually with some post-doctoral fellows, and even a few visiting professors from Europe. We learned a lot from each other.

In addition to teaching the graduate Physiology course, I soon launched my new course, *Environmental Physiology,* which included a laboratory several times a week. I also lectured in the Medical

Physiology course, and was about to be given additional mentoring responsibilities with these young medical students.

**In Iowa laboratory with tame Arctic Ground Squirrel.**

During the decade from 1953 to 1963 there was a sense of dynamic progress in medical education in the physiology department. Previously, medical school classes, made up of about 440 students, had for the most part been kept together. But a new system of teaching was being introduced. Professor Hines, the department head, divided the medical class into groups, or sections, of 30 students. Each physiology faculty member was responsible for at least one of these sections. Each section was to do a research project, or a specific review of the literature. Surprisingly, the research results were often good enough to publish.

After a while, Hines changed the system further, dividing the medical class into even smaller groups. Each member of the department was to take three groups of eight students and work with each group to complete a research project. The students were all freshman in medical school and most had not rolled up their sleeves and gone deeply into a research project before. In my laboratory, we studied the comparative physiology of biological clocks in various animal species. We also did some work on biological clocks and temperature regulation with human subjects. Two of my medical students wrote a technical paper about their research that was published. From my viewpoint, the activity associated with these groups of medical students was totally successful. It was a smart move for Hines to break the large class into small groups.

When I first came to this big university, I had been ready for the rule one often hears about: "Publish or Perish." But I was rather unprepared to realize that one must travel in order to verbally present one's work. Fortunately, travel was supported by our department. There were two organizations where we presented much of our research: the American Physiological Society (APS) and the International Society for the Study of Mammalian Hibernation. The APS met in various cities, and we were expected to attend these meeting to give talks about our research and results from laboratories and graduate students. Attending these meetings is useful for making contacts with others in your field, for learning what others are doing, and for fostering collaborations.

This meant that I traveled quite often to various cities and to Europe to take part in symposia and to present work from my laboratory. As a single father with two children, I often found it difficult to be away so much, but with the help of friends and babysitters, I managed to go as often as was needed.

Of the many activities, it is impossible to say which I enjoyed the most, but with such a busy schedule it was extremely helpful to have a twelve-month appointment after being at a small liberal arts college that had provided a salary for only nine months!

## Mary

When I had been interviewed in Iowa for this position, to tempt me to accept, I had been told about a talented artist who illustrated technical papers for the medical faculty. Upon arriving in Iowa City, I soon needed illustrations for my publications and so I met the young artist, Mary Arp. Mary and I had much in common and became friends who shared a love of nature, science and horses—Mary even had a mare, Contessa, who lived nearby. When my marital status changed to "single," our friendship led quite naturally to frequent dating, and it wasn't long before we became engaged.

**Mary on Contessa.**

Mary's family lived in the Quad Cities about 60 miles away. Fortunately, they approved of our love, and gave their support to our engagement. Mary's mother insisted we give her a year to plan a large, high-society wedding; Mary was impatient, but we acquiesced to her mother's wishes. I was unaware that my future mother-in-law had placed an engagement announcement in the local newspaper until my students began to congratulate me!

In 1956, we were married in a large Methodist Church in the Quad Cities, with my minister father officiating. I was 41 and Mary was 27. Kit had become a young man of 20, with a job and apartment, and Victoria was a teenager, who at the time lived at her mother's. Mary and I spent most of 1957 abroad, as I had a Fulbright Fellowship in London—we always thought of that year as our honeymoon.

Mary's sister Josephine once said she was sorry for Mary because she was now locked into a traditional role with all the social duties of the wife of a medical school faculty member. But this turned out to be far from the case.

Mary eventually became an active member of my research team and was often a co-author on scientific papers. It is true that she illustrated our papers, and then served as editor, but soon she was much more, taking an active part in our breakthrough work, which was to implant physiological radio transmitters in large animals, even administering anesthesia to bears and other wild animals. She became a colleague, working at my side in the field under all sorts of conditions.

Mary liked working with animals and with her husband, and she preferred to spend as much time as possible in wild places with a mountain breeze in her face.

## Christopher

Although I had been Christopher's step-father and had raised him from age four, our relationship had never been formalized. Thus, Mary and I officially adopted him. He was an adult by then, and would keep his last name, but it was a meaningful event for all of us. Kit was an undergraduate at the University of Iowa, working toward a double major in biology and foreign language (German and Russian). He was an excellent linguist—one of those people who seem to just absorb languages without having to study. His good ear for languages also made him an excellent mimic.

As always, Kit couldn't stay away from the outdoors, and we spent a lot of time roaming the woods together. He soon became one of the best herpetologists in Iowa. While he was an undergraduate, he discovered a new species of reptile not previously known in Iowa: a musk turtle (*Sternotherus odoratus)* (Dodge 1956), which he found in a spring-fed, cold pond overgrown with watercress. Later, he found yet another species of turtle new to Iowa. He was good at finding rare species.

In an incredible coincidence, one day while this expert on reptiles was hiking along the bluffs of the Mississippi River, he sat down against a tree to eat his lunch and put his hand on a rattlesnake, which bit him. He finished his lunch, hiked out, and drove himself to the hospital, which then had to fly in the antivenin from another state while his arm swelled to twice its normal size. When Chris recovered, he asked me to help him write a technical paper on the snakebite and rattlesnakes in Iowa (Dodge and Folk 1960).

Soon after the paper was published, Dr. Hines, our crusty but loveable, and often funny department head, came to my office. He stuck his head in my door, leaned against the door jam, and said,

"Folk, I don't know why I hired you!" Shaking his head, he muttered, "Anybody who would let his son get bitten by a rattlesnake—just so he could publish a paper!" More head shaking.

In all, I was fortunate to publish four papers with my son Chris.

## Victoria

Victoria too had a brilliant and retentive mind, but perhaps because she saw that her father worked all the time, she was not inclined to a career in academia. She attended a public high school until I found out that at home, she could "study" while simultaneously listening to television and still get all A's, which exempted her from having to take final exams. I felt she needed a more rigorous scholastic challenge. I took her out of the public school and put her in the University High School. There she got mostly B's.

Then I learned that Victoria, like many other teenagers, was spending too much time in the soda shop at the local drug store. I decided that she needed another hobby, something to keep her occupied, preferably something outdoors. Horses would be the answer, or so I thought.

We already had one horse, Mary's mare Contessa, but I knew Vic wouldn't ride alone, so I bought another horse and boarded them at a farm just outside of town. I tried to get Vic involved in Pony Club, which was run by a former Olympian rider, Betsy Coester. But I think probably I was more interested in that opportunity than she was—I know I spent many hours at the Coesters' farm cleaning stalls for the Pony Club! Occasionally, Vic and one of her girlfriends would ride through City Park, and afterwards talk about all the boys they had seen, and more importantly, which boys had seen *them*.

I took Victoria with me to a national meeting in California, hoping to broaden her horizons. Instead—and perhaps due to a lack of maternal guidance—during her last year in high school she eloped with the son of another professor. After their marriage by a justice of the peace, she went through those marriage vows again with the young man, but this time in the Episcopal Church, at my insistence.

Afterwards there were years when Victoria had little to do with her father, which was a very unhappy part of my life. I realized sadly how her mother's absences and my work schedule must have contributed to make Victoria feel deprived of family love while growing up, despite my always having loved her very much.

Being married and having a family didn't slow Victoria down: she completed college in three years with a degree in Art and French. She took graduate courses in education and journalism, and taught French,

English and Drama in high school. She went on to write two books, and conceived and published the first woman's newspaper in the United States, which is now housed in the Library of Congress. Then, at age 45, Victoria went to law school and earned her JD degree. She still works in the law profession.

Happily, we have again grown very close. I feel fortunate to have such a beautiful, accomplished daughter in my life, as well as her wonderful children—Jennifer, Emily and Joshua—and their eight children.

## Environmental Physiology: A New Discipline Emerges

Definition: *Study of the reactions of animals, plants and humans to hostile environments such as heat, cold, altitude, high pressure and extremes of light exposure.* The sum of these reactions in each environment is called *acclimatization.*

This discipline accepts, describes and analyzes the state of complex interaction between mammalian life and the physical environment. Thus, when studying an animal from this perspective we can never isolate it from its physical environment, nor can we move it without taking its environment with it. Part of this overall discipline is the study of Darwinian genetic adaptations.

Humans have always been aware of environmental influences on living organisms. Hippocrates investigated the effects of weather and climate on humans from a medical viewpoint. Darwin's theoretical view stressed the interrelations of life and the physical environment.

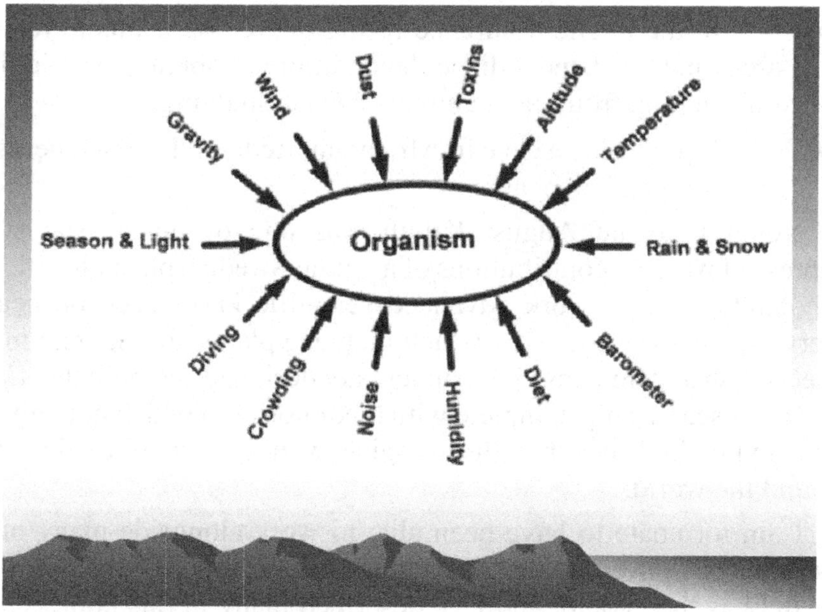


**Diagram illustrating some of the environmental factors that impact a living organism. We might also consider radio waves from cell phones, data overload, and other technological issues that have become so much a part of our environment. (Source: Folk, Riedesel and Thrift 1998.)**

Although environmental physiologists conduct many experiments under controlled laboratory conditions, of necessity they also often work in the field under the same hostile environments whose effects on living organisms they are studying. Indeed, many of these intrepid scientists have used themselves as subjects in these unfriendly environments. I've given many examples in this book, but there are others worth mentioning.

- In 1775, Blagden and Fordyce took a dog and some obliging associates into a room heated to 198°F; they carried the dog in a basket so its feet would not be burned.
- Hammel and colleagues immersed their own feet and legs in painfully cold ice baths to compare their reactions with the indigenous people at the tip of South America.
- In 1902, Danish physiologist August Krogh took his instruments to small villages in Greenland to measure the influence of a high fat diet on the metabolism of the Eskimos.

- In the Arctic, Laurence Irving drove dog teams to remote areas and lived under difficult conditions to obtain physiological information from the mammals he was studying.
- Deep inside a cave in Africa, Ian Redmond studied herds of "salt-mining" elephants.

Nobel Laureate August Krogh was one of many who were impressed with the contributions of a young Swedish physiologist, Per Scholander, whose work advanced scientific knowledge on nearly every organism he studied—which included plants, diving mammals, insects, fish and humans. Scholander later designed and built the *Alpha Helix*, a research ship complete with laboratories in which not only he, but also hundreds of other physiologists would carry out field studies around the world.

I am fortunate to have been able to work alongside many other distinguished physiologists who chose to do as much of their work in the field as they could, and I've described many in this book. These include Bruce Dill, Donald Griffin and Herman Rahn to name but a few.

### *Development and Name*

During my last year at Bowdoin, I had been putting together a new course I called *Environmental Physiology*. I had carried all my notes to Iowa, and after a while, began to teach this new course.

As I developed the course, I realized that a central theme to studying responses to extreme environments was *acclimatization*. One could look for this phenomenon under the extreme environments of heat, cold, high altitude, high pressure, and also within the study of biological clocks—a topic also influenced by the outside environment.

Because this was an emerging field that had never formally been named, I realized that I would need to select a label for it. I rejected the term *Adaptational Biology*, although this would have been partially satisfactory because of its connotation of evolutionary change. But it did not go far enough.

I was reluctant to use the term preferred by Per Scholander, *Expeditionary Physiology,* because I knew many academicians who felt that the title of *Expeditionary Physiologist* would not be appropriate for a position in a university or medical school. I remember discussing this with Scholander back in the mid-1940s at the Fatigue Lab, where I had hesitantly mentioned the possible drawbacks of that title, and had tentatively suggested *Environmental Physiology*. But Scholander had obviously liked his title better!

Two decades after those discussions with Scholander, I was teaching my new Environmental Physiology course in Iowa, and it became clear to me that *Environmental Physiology* was indeed the right name for the field. I wrote my first textbook, *Introduction to Environmental Physiology*, while on a working sabbatical in Fairbanks, Alaska in 1964-65. It was published in 1966 by Lea and Febiger; I still have a copy of the first edition, which the publisher prepared for me with gold-leaf edging. That book, and its second edition, *Textbook of Environmental Physiology* (1974), were family affairs: Mary did the illustrations and helped with all the editing.

To our knowledge, the 1966 textbook was the first ever published that used the term *Environmental Physiology*. At some point I heard from a student that a professor in Canada had published a lab manual with that term in its title. As to whether he or I first published or used the words in the classroom is actually irrelevant.

From 1966 to '68, much to my surprise, seven other books were published in England and the United States with *Environmental Physiology* in their titles. I can only interpret this outburst of books on this topic as acceptance of the area as a new discipline. Obviously the discipline itself was booming. Another step in accepting the discipline was the appointment of my old boss, Woody Belding, as Professor of Environmental Physiology at the University of Pittsburgh.

Looking back, I can see that it was a time for the development of new disciplines or areas within an established field. Within the field of ecology, a similar dilemma resulted in the demarcation of a discipline called *Physiological Ecology*, a perfectly satisfactory discipline to be found in a biology department. Another field developing at that time was *Biometeorology*. An international organization formed with this name is flourishing today with numerous members and a most successful journal.

## Excursions into the Outside World

The students working with me were always involved with either biological clocks or the study of mammalian hibernation. Much of our work was done in the laboratory on white rats, hamsters and ground squirrels.

The ground squirrels gave some respite from being restricted to the four walls of laboratories and lecture rooms because I needed to capture specimens for my studies of hibernation. Also, two of my PhD students elected to study the biochemistry of hibernation using cave bats as models. This of course meant that we had to go to the caves where bats hibernated to obtain them. I was pleased that there were occasional ventures into the outside world to do science.

Meanwhile, some of my friends from graduate school were leading a different kind of life. Peter Morrison, for example, spent part of his time in the field in Australia studying the physiology of marsupials, while also traveling to South America to study the physiology of the guanaco. I read of their experiments with interest, envied their outdoor adventures, and thought wistfully of their work as excursions into the private lives of unusual animal species.

As it turned out, in addition to my faculty job at the University of Iowa, I too would be able to carry my research to the far horizons—to the shore of the Arctic Ocean. Essentially I was to invade the territory of the wolverine, the polar bear, wolf and the tundra grizzly.

# Chapter Nine

## Studying Animals and People in a Harsh Environment: Alaska
**1961–1977**

FOR SIXTEEN YEARS, in addition to teaching and doing research in Iowa, I maintained a laboratory at the tip of North America at Point Barrow, Alaska—one of the most unusual and extreme environments on earth. My earlier investigations into biological clocks had been done in the laboratory under artificial light and darkness, but I had longed to study biological clocks in the outdoors, in a locale where extreme light conditions were normal and seasonal. With its several months of continuous natural light in the summer and several months of continuous darkness in winter, Point Barrow was the perfect place.

I had been encouraged to go to Point Barrow by a colleague at Iowa, Alton "A.K." Fisher, a pathology professor at the UI College of Dentistry with whom I had studied the dental tissue of ground squirrels. A.K. was also an Admiral in the Navy and had worked in the Naval Arctic Research Laboratory at Point Barrow to study gingival tissue of seals. I had many questions about establishing a laboratory in Alaska, not the least of which was, "But can I take my wife?" "Of course," he replied, "I took mine!"

I received a grant from the National Science Foundation to support this work, and it was renewed continuously for 16 years. Mary not only went with me, but also worked alongside me on every project. Over the years, I also took seventeen students to Alaska for various periods; a couple of them lived at Point Barrow for two years while collecting data for their PhD theses, which they finished back in Iowa.

## Naval Arctic Research Laboratory

Even before World War II, Laurence Irving and Per Scholander had discussed the need for a permanent research laboratory in the Arctic. In 1947, with sponsorship from Swarthmore College and the newly formed Office of Naval Research (ONR), Irving and Scholander established what would become the Naval Arctic Research Laboratory (NARL) at Point Barrow, Alaska. This laboratory was on the Arctic Ocean shore, and five miles from the Iñupiat village of Barrow.

**Small section of NARL as seen from the air.**

Using Naval Base facilities that had been set up at Point Barrow for oil exploration in one of the Navy's Petroleum Reserves, the NARL became an elaborate complex with individual small laboratories for visiting scientists. During the years I was there, the "spark-plug" of this scientific activity was Max Brewer, its director.

Other scientific programs operated out of the NARL, including the famous *ice islands* or "Drift Stations." These were research stations set up on huge chunks of ice (averaging 1.5 miles wide and 70 feet deep), which had broken off from ice shelves and were floating in the Arctic Ocean. One of the ice islands, ARLIS II, supported many different research projects in 14 pre-fab buildings over its four-year operation. It had to be abandoned however, as it was eventually carried north by the currents across the Arctic Ocean, too far from Point Barrow for resupply.

When the Navy's oil operations moved to Prudhoe Bay, the Air Force took over the management of the base as part of their Distant Early Warning Line ("DEW Line"), a series of defense radars set up across the far northern borders of the North American continent to detect incoming missiles or ground forces during the Cold War. The NARL continued to be sponsored by the Office of Naval Research.

Two years after he established NARL, Lawrence Irving moved to Anchorage, Alaska to head the physiology section at the new Arctic Health Research Center. In 1962 he moved again, this time to the University of Alaska Fairbanks where he established the Institute of Arctic Research, which he directed until his retirement in 1966. I visited with him often when I traveled through Fairbanks.

## Life in the Far North

The Arctic Circle has been called "a geographic ring crowning the globe," with the North Pole at its center. The outer line of the Arctic Circle begins about 1700 miles south of the Pole, about 25 miles north of Fairbanks, Alaska. This imaginary line marks the southernmost latitude (66°32' N) from which you will not see the sun rise above the horizon on the year's shortest day (winter solstice), nor see it set on the longest day (summer solstice). As you head north into the Arctic Circle, you will notice that continuous daylight and continuous darkness last much longer, increasing from one day near Fairbanks, to a full six months at the Pole.

Point Barrow, Alaska, at Lat 72° N, lies well within the Arctic Circle, being approximately 500 miles north of Fairbanks and about 1200 miles south of the Pole. At Barrow, the sun circles above the horizon from mid-May to early August, but does not rise above it from mid-November to late-January. The polar twilight lasts several months longer on either side of these extremes. The cold is also intense, with winter temperatures sometimes dropping to -56°F, and for 10 months of the year, the daily minimum temperature is below freezing. During Barrow's brief summer, the thermometer may occasionally reach as high as 78°F, but the average is just 40°F.

**Arctic Region showing the Arctic Circle (dotted line). Barrow is indicated by a star within circle at 11 o'clock. (Source: *The World Factbook 2009.* Washington, DC: CIA.)**

For the first several years that I was going to Point Barrow, I felt I was living a Jekyll and Hyde existence, traveling in the space of 48 hours from the active, tree-shaded college town of Iowa City to the near-empty vastness at the tip of North America. The whole place felt isolated and primitive.

A person standing at the south end of the laboratory looks out over 200 miles of treeless flat tundra, broken in the far distance by the snowy peaks of the Brooks Range. That flatness is deceptive, however, as the land actually slopes imperceptibly upward from the Arctic Ocean to the foot of the mountains. The Brooks Range is the farthest north Continental Divide, separating not only vast watersheds, but also Alaska's interior and Arctic zones. Thus, the entire tundra north of the Brooks Range is known as its "North Slope."

The remote and magnificent Brooks Range is reachable only by dogsled in the winter, by river in the brief summer, or by airplane. Until recently, it was thought that this mountain range was the northern beginning of the Rockies, but geologists have determined that these mountains rose perhaps one hundred million years *earlier* and are not related structurally to the Rockies—only by their splendor. The highest peaks in the Brooks Range stand at about 10,000 ft, with the average being about 6,000 ft.

The North Slope is all permafrost, which means that the subsoil stays permanently frozen and only the top two feet of ground ever thaws. From October through May, the tundra is covered with snow. In spring however, because melting snow cannot penetrate permafrost, the North Slope becomes a 60,000 square-mile marshland, dotted with lakes, ponds, streams, and puddles that, consequently attract hordes of mosquitos. Taking a walk across the thawed land requires hopping in a zigzag fashion from one dry piece of tundra to another. But the tundra also blooms in summer as wildflowers awake from their winter sleep, spreading a blanket of beautiful colors. On our zigzagging walk we find moss, lichens, short grasses, heather and occasionally, a dwarf willow. It might seem surprising that this area is defined as a "polar desert," but rainfall is scarce, and average annual snowfall is only 30 inches.

It was a remarkable experience to see the sun overhead, going in a circle above the horizon all 24 hours as the summer developed, and to watch this circle of light gradually broken in the east as the late season developed. Eventually there was no sun above the horizon and only a moving burst of light below the horizon to represent the true sunrise beyond the Brooks Range. The continuous light of summer was soon to be replaced by the continuous darkness of winter.

Unless one traveled by dogsled (the nearest town was 200 miles away) or by ship during the brief summer when there was no sea ice, the only way to reach Barrow was by airplane from Fairbanks or

Anchorage. Planes landed near the Naval Base on runways made of metal grid-like mats laid atop gravel-packed snow, or in summer the mud. NARL had five DC-3 cargo transport planes—sturdy 2-engine propeller aircraft left over from WWII. On those flights to Pt. Barrow, passengers were stowed amongst the cargo; I remember one flight where I sat atop a crate of celery!

The first time I landed there, in 1961, as I was being conveyed from the airplane to the lab, I saw a sight that was an archetype for this far northern land. Across the unbroken expanse of the frozen Arctic Ocean was an Eskimo kneeling on a sled that was being pulled by a team of dogs. I stopped the jeep and jumped out to take a picture (see center photo on cover). I was struck by the realization that this scene would have looked just the same 100 years earlier. Until, that is, I turned back to the jeep and saw the base spread out behind me.

Mary and I lived on the base in a small apartment comprising one-half of a Quonset hut, as did most other visiting researchers and their families. The "hut" was furnished and came equipped with blackout curtains to block the midnight sun. It was heated by oil brought in big barrels. The little place was cozy, and during our long visits we added our own touches with fur pelts and ivory which we received as gifts or purchased from the Iñupiat.

**At dinner inside our Quonset hut.**

**"265 Ocean Blvd." One of the Quonsets where Mary and I lived for up to 6 months in Pt. Barrow.**

Living in barracks on the shoreline of the Arctic Ocean was tremendously exciting. Snow piled up to the top of the Quonset huts during winter storms. In summer, we could sit on our tiny porch to look out at the Arctic Ocean and watch the Eskimos in their skin boats. Everybody wore a winter parka, even in the summer (sometimes unzipped) because the breeze off the ice pack was cold. But we were lucky to have that breeze, because if we went very far towards the mountains we would be bitten to death by the mosquitos (except for repellants). In the middle of summer there were sometimes small snowstorms, and afterwards we would occasionally find polar bear tracks on the porch of our Quonset hut! To my knowledge, the bears never tried to break into the huts.

Nature could wreak havoc at Point Barrow in a variety of ways, with winter storms, fog, strong winds, and wind-driven waves. In October of 1963, while I was safe in my Iowa City lab in the lower 48, nature unleashed a storm on Barrow so severe that seawater as deep as three and a half feet swept through NARL, moving buildings, flooding the freshwater lake, even killing animals. The current was so swift that tractors, two-track vehicles and trucks were useless. Thankfully, most of the animals were saved before the tractor rescuing them was finally sunk. I was saddened to learn that several of my animals had drowned—one wolf, two wolverines, and three foxes. Fortunately, there were no human deaths, but one of my graduate students living in Barrow at the time, Gary Shook, stood on his kitchen stove with his wife and two children to avoid the inundation. Some people were evacuated, while others made valiant efforts to save the animals.

**Gary Shook and Roger Hedge in front of a sign welcoming people to Barrow. The sign is made of the whale's scapula, which is suspended between two ribs. On the ground below is a whale skull surrounded by vertebrae.**

## *Oil Changes the Landscape*

After oil was discovered at Barrow more people came, which resulted in Barrow building a small commercial airport near the village. We flew into that airport from then on, instead of the more primitive one on the base. Ground transportation, however, entailed a four-mile ride to and from the airport through all kinds of weather: snow and bitter cold in winter, or the ubiquitous mud of Arctic spring. Barrow's "boom" lasted only until 1968, when oil was discovered in Prudhoe Bay, about 200 miles to the east.

To move the crude oil south from the huge oilfields in Prudhoe (32 billion barrels in 1968), the giant 800-mile Trans-Alaska Pipeline was built. Next, a 414-mile gravel road, named the Dalton Highway but dubbed the "Haul Road," was laid parallel to the pipeline to bring supplies from Fairbanks to Prudhoe.

The road didn't change the way we got to the lab at Pt. Barrow for several reasons. First, it was a long, slow drive on the Haul Road from Fairbanks to the Arctic shore. And second, Prudhoe was still hundreds of miles from Barrow and there were no roads or flights between those two places at that time. (Now there are some flights between those far northern cities, though it would be a slow, roundabout way to reach Barrow.)

But I wanted to see this great undertaking of a road across the tundra. So one day, Mary and I and a student spent several hours driving the Haul Road north from where it begins about 70 miles outside of Fairbanks. During the relatively short stretch we covered, there was no dramatic scenery to speak of, just marsh on both sides, and of course a huge pipeline looming on supports above the ground running along beside the road. Farther north, we knew, there would be stunning views as the road crosses the Brooks Range.

The gravel road was full of bumps made by both the weather and the heavy haulers that drive to and from Prudhoe. We saw tremendous scrapers with giant wheels trying to repair and smooth the road. Even today, more than 40 years later, the road is still largely gravel, and it remains the only road that ventures into Northern Alaska, crosses the Brooks Range and North Slope, and ends at the Arctic Ocean!

As for reaching Barrow today, it is still the case that the only way to get there is by airplane, dogsled, snowmobile, or—during the three to four summer months when the sea ice is melted—boat.

## The Eskimos

Most of Alaska's indigenous people are collectively referred to as *Eskimos*; they often use this term when referring to themselves and apparently do not feel it to be pejorative. (This is different in Canada, where the term *Inuit* is strongly preferred.) In Alaska there are three ethnic subgroups distinguished by language and location: in the far north along the coast of the Arctic Ocean live the Inuit (Iñupiat); in western areas and south along the Gulf of Alaska live the Yup'ik, and from the Aleutian Islands come the Aleuts. In Alaska, *Inuit* and *Eskimo* are often used interchangeably, however, the Yup'ik are understandably not pleased about that, because they are not Inuit.

Ethnicity, origin and relationships among the aboriginal people of the Arctic (which includes Alaska, Canada, Greenland and Siberia) are complex, and even Eskimo scholars may offer differing explanations. Therefore, my humble attempt to briefly clarify this fascinating subject must suffice.

## *The Iñupiat*

The Iñupiat Eskimo village of Barrow is ten miles west of the Point Barrow headland, and four miles west of NARL. It is the farthest-north American city. It was named after Englishman Sir John Barrow, who championed Arctic exploration, but in the Iñupiat language, the village is *Ukpeagvik,* which translates roughly to "place where snowy owls are hunted." Barrow is one of the largest Eskimo communities in the world; its population was 3,000 when I was there in the 1960's, and has grown to over 4,400 today. A lot of snowy owls also live there!

**Eskimo children in Barrow, Alaska posing for my camera.**

The Iñupiat have lived in this area since 500 A.D. according to archeologists. In fact, so many unmarked shallow burial sites containing ancestors from pre-contact times have been found in the tundra around Barrow that researchers must be warned not to disturb them. Christianity was brought to the Eskimo a century ago, but as with many indigenous people so converted, their beliefs incorporate both the old and new.

**Hunters returning from hunt on ice pack.**

On the Arctic shore, the people of Barrow still live by fishing and hunting, although to a lesser degree today than in the past. Depending on the time of year, they hunt whale, seal, polar bear, walrus, waterfowl, and caribou. Shrinking sea ice caused by global warming has become a serious threat to the Iñupiat as well as to the polar bear. Both need extensive strong sea ice in order to get out into the ocean where hunting and fishing is best. The old-timers at Barrow see the effects of the melting Arctic ice pack firsthand, reporting that the "July of Ice" is becoming a "July of Water."

**Drying caribou meat. Street scene, Barrow, AK.**

Somewhat similarly to the way I'd once observed the military changing from horses to machines, I also witnessed the Eskimo change their mode of transport. During my first decade or so in Barrow, I saw the Iñupiat traveling exclusively by dog team, but sometime in the '70s, the snowmobile came on the scene. Over the years I saw more and more villagers adopting the snowmobile as their primary means of getting from one place to another.

### *Whaling*

"The giant bowhead whales live at least two human lifetimes." This is what the Iñupiat have always told their children, and now, the longevity of the bowhead has been confirmed by dating artifacts, such as old harpoon heads, found in whales. With a lifespan of more than 150 years, bowhead whales are the longest-living mammals on earth.

Eskimos are the only people who can still legally hunt bowhead whales. Out on the ocean in their skin boats, they look for a whale who will "offer itself to a worthy hunter." After a successful hunt, prayers of thanks are offered, not only to the Almighty, but also to the whale.

**Boy in skin boat, or *umiak*. (Photo: USGS.)**

With my many sojourns to Barrow, it was inevitable that I would observe the processing of a whale. As a biologist, I wanted to learn more about whale anatomy as well as to observe native customs.

The residents of the village would first combine their muscle power to pull the leviathan onto the ice. Then they cut up the whale using hand tools. Eskimo children ate slices of the blubber like it was candy. Taking a whale was always a big event, and everyone in the village shared in the feast and received a portion of the meat.

**Skin kayak (umiak) on the street of Barrow Village.**

I noted that when I was near the whale, I found myself giving in to a human response, which was to rub the side of the whale to experience what it felt like. This whale was a female, and because I am interested in endocrinology, I asked to hold the ovary; it seemed to weigh about half a pound. My mind reeling, I asked myself whether it was bigger or smaller than I'd expected. It seemed small, I decided, and realized it was probably because it was not their breeding season. The size of the ovary would vary during the two years between calves, becoming a much larger organ in the breeding season. This is an example of *the economy of nature* which provides a surplus of ova.

I also held the whale's eye, which fit in my hand. It was about the size of my wrist watch, and I was struck by its small size relative to the whale's large skull. Here is an example of the way *anatomy correlates with function*: the wall of the eye must be thick to protect against the crushing pressure of water during deep dives; if the size of the eye were relative to the size of the skull, it would take up too much room. This whale, a bowhead, was a surface feeder, but in a deep diver like the sperm whale, which can dive to 5,000 ft, the eye is even smaller. Seeing my interest, the hunter gave me the whale's eye, which I took back to Iowa preserved in formaldehyde as a gift for Ralph Janes, my good friend who studied the fundus of the eye.

## *Native Naturalists*

Quite a few Barrow residents worked at the Naval Base and NARL as carpenters, machine operators, mechanics, guides, dog drivers, trappers, cooks and office support. In addition, many were expert scientific observers. Max Brewer, a long-time director of NARL, gave the Iñupiat the credit they deserved, writing, *It is not necessary to have formally passed through the halls of a university to become a natural scientist... The very existence of the Iñupiat was often dependent on keen observation and memory. Thus their very existence in this harsh environment made them great naturalists (Brewer and Schindler 1997).*

**Kenny Toovak, our NARL "go-to" man.**

It would not be an overstatement to say that the success of my experiments was in the hands of several capable Iñupiat, particularly Kenny Toovak and Pete Sovalik. These two men were leaders in their community and famous for their contributions to Arctic exploration before and during WWII.

**Pete Sovalik with wolf cubs.**

Kenny Toovak was operations manager of NARL. Kenny was NARL's "go to" man who could find a way to make anything happen, from building cages strong enough to hold 1,000-lb bears to taking visitors out in dogsleds to see the tundra. Kenny's contributions to science at NARL were enormous, and I felt honored to be present several decades later when he was awarded an Honorary Doctorate from the University of Alaska. Today, Kenny is still facilitating international scientific research as a board member of the Barrow Arctic Science Consortium.

Pete Sovalik was the Iñupiat I worked with most closely because he helped care for my animals every day, year round. These animals ranged from lemmings and marmots to wolves and bears. He contributed greatly to our program, capturing the first lemmings when we started our colony, bringing in our first wolf cubs, and helping move the huge bears after they'd been tranquilized. I remember Max Brewer telling me once, never to go into the wolves' enclosure without Pete. That went double for the bears! Pete had a way with animals, and enjoyed the work.

Pete's excellent memory and contributions to NARL's research were widely acknowledged, and Max Brewer credited Pete with "training a whole generation of young scientists in how to observe and interpret nature, and how to survive in the Arctic" (1997).

### *Nunamiut at Anaktuvuk Pass*

Anaktuvuk Pass is an isolated, stunningly beautiful mountain pass in the Brooks Range. Its lowest elevation is 2,200 ft, with nearby peaks rising steeply to 6,000 ft. The Pass is about 250 miles south of Barrow across the tundra, and a similar distance north of Fairbanks. It is reachable only by dogsled, snowmobile or airplane.

A regular migratory route for caribou and birds, the area has become home to about 300 Nunamiut Eskimos. These inland-dwelling Inuit were once nomadic, following the caribou, whereas most of Alaska's Inuit are coastal people whose lives have always been centered on the sea.

**Simon Paneak of Anaktuvak Pass and Pete Sovalik practicing their dance moves.**

While I was in Alaska in the 1960s and '70s, the Anaktuvuk Pass area was home to only about 100 inhabitants. Best known to those of us at NARL was Simon Paneak, a highly respected native naturalist and colleague of Larry Irving. Paneak was a village elder and also a cousin of Pete Sovalik. I visited the Pass several times to take photos of this magnificent area and occasionally to research wildlife.

The Nunamiut of Anaktuvuk Pass have become famous for developing a unique art form, the "modern-day" Eskimo masks. Although the Iñupiat had once used ancient ceremonial masks, their use had died out after the people became Christians, and few, if any, of the elders even remembered the old masks. The masks seen today, known as "Anaktuvuk Masks," were not made with thoughts of evoking a bygone era, nor of creating a new business, yet that is exactly what has happened.

The story of the first Anaktuvuk masks is interesting in part because of the subtle interaction between two very different cultures. I heard the story from Larry Irving, who played a small part in it, and I will relate it here.

In 1951, two Nunamiut men on a hunting trip were discussing the upcoming Christmas dance in their village. The celebration was a time of playfulness and merriment, and the two men were wondering what they could do to surprise their friends and relatives at the dance. One of the hunters talked about seeing some Halloween masks at a store in Fairbanks, and as neither Halloween nor its customs were known in the village, the men hit upon the idea of making masks to wear to the dance. The masks they created were primitive and made from materials at hand, primarily caribou skin and sinew. The hunters cut holes for eyes and mouth, sewed on a nose, fashioned hair, and added a fur ruff to look like the ubiquitous parka hood. It took almost two days.

When the two men wore their masks to the dance, they were a huge success, surprising and delighting everyone. Afterwards the masks were put away and nearly forgotten.

A bit later, when Larry Irving was visiting his friend and co-worker Simon Paneak at Anaktuvuk Pass, Simon showed Larry the masks and related the story of the two hunters. Appreciating their uniqueness, Larry Irving offered to buy the masks, and proudly displayed them in his home in Fairbanks. As other outsiders learned about these distinctive masks, they urged the Inuit to make more of them for trade, *et voila!*

**Anaktuvak Masks, highly prized cultural icons.**

Designing and making masks became a cottage industry in Anaktuvuk, and many of these unique masks now hang in museums and private collections. While the original masks were not made to restore lost traditions or to sell, Anaktuvuk Masks are now in demand throughout Alaska, and are highly prized as cultural icons.

## Research

### *Biological Clocks*

My PhD thesis had been on the topic of the biological clock of the white rat under constant daylight. The study of biological clocks—also known as *circadian rhythms*—was a reasonable topic because thousands of humans and their livestock around the globe experience 24 hours of continuous daylight each summer. The great scientist, Nathaniel Kleitman, first gave his attention to this lifestyle in his 552-page book entitled, "Sleep and Wakefulness."

Initially, my studies in Point Barrow would focus on the circadian physiological clock, which governs the daily behavior of humans and animals. In the Arctic summer, the clock is dominated by continuous daylight, but later is subjected to months of continuous darkness. During this period, daily activity for people and animals in the free-environment is influenced by their own innate biological clocks. These clocks are sometimes quite different from animal to animal. For example, some animals are *nocturnal* (foxes), while others are *day active* (caribou).

The NARL director, Max Brewer, was deeply interested in our investigations, as was the director of the ONR in Washington, DC, Max Britton. They liked the unusual method I had developed of studying the physiology of local animals by implanted radio telemetry, and wanted me to expand my project to study all Arctic mammals including the three species of bears—black, grizzly and polar—which I eventually did.

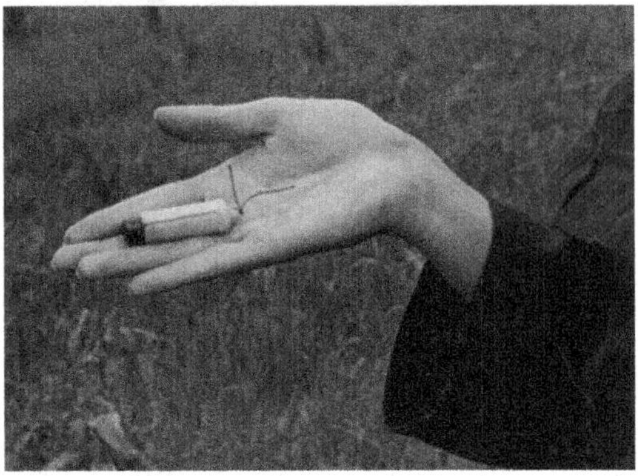

**One of our radio transmitters.**

We were the first to do this recording of physiology by implant telemetry. Our implantable transmitter had been developed in Iowa by my brilliant student, Warren Essler, and we named it the Iowa Implantable Radio Transmitter. There were models weighing from 10 to 20 grams for various size animals. The method of recording physiology by radio transmitter could be done by either attaching the radio on the surface of the animal or person, or by opening the abdominal cavity of the animal and sewing the transmitter inside the body.

I established the surgical technique for implanting these radio transmitters in the animals' peritoneal cavities. By this unusual method, we could record the animals' physiology over days or months, something we knew would be useful, especially in very large animals and when animals were experiencing no sunlight for as long as six months.

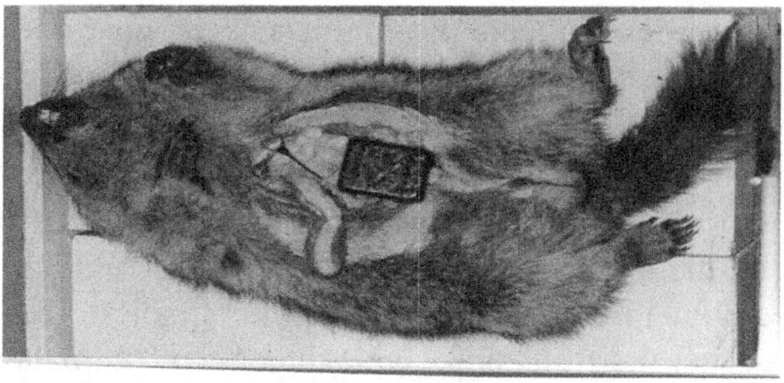

**Implanted radio transmitter in Arctic Ground Squirrel. Later transmitters were 1/5 the size of this one.**

In all, over many years in both Barrow and my Iowa City laboratories, we studied the physiology of twenty-one species of animals.

| *Species* | *# studied* |
|---|---|
| Macaque monkey | 2 |
| Bat | 1 |
| Cat | 3 |
| Dog | 2 |
| Guinea pig | 1 |
| Hedgehog | 1 |
| Lynx | 1 |
| Snowshoe Rabbit | 9 |
| Domestic rabbit | 3 |
| Seal | 1 |
| Porcupine | 6 |
| Wolverine | 3 |
| Arctic ground squirrel | 4 |
| Woodchuck | 1 |
| Raccoon | 4 |
| Black bear | 3 |
| Grizzly bear | 3 |
| Polar bear | 2 |
| Arctic marmot | 5 |
| Arctic wolf | 4 |
| Arctic fox | 10 |
| Reindeer | 1 |

Except for the reindeer, the animals were instrumented in the peritoneal cavity with a radio transmitter to record body temperatures, heart rates and EKG, which captured their daily circadian rhythms of sleep and wakefulness.

It was a fortunate thing that, whether in the laboratory or in the free environment, the mammals "graciously accepted" in their abdominal cavities, without infection or rejection, the small sterile radios that would broadcast their physiological state. Some animals went on to live long lives and have healthy offspring afterwards.

In our program to compare the 24-hour biological clocks in animals and humans, we began within the normal environment of daylight and darkness, then moved to the responses of the clock to 82 days of continuous light and 82 days of continuous darkness. In people, of course, we used observations, not transmitters. We watched and wrote down what was happening in the village every hour of every day for one week. We did this in Barrow and another village we felt had not been as westernized.

We found differences between the visiting scientists and the natives. The visitors generally kept to their temperate zone sleep habits, using black curtains to block out the continuous light. Eskimo villagers on the other hand, were often up all night. At midnight, men hunted and children played in the streets. I'm sure they slept but I don't know when. I asked more than one Eskimo hunter, "When are you going to sleep?" They would smile broadly and say "When it gets dark!"

Although our students were not subjects in that study, we noted that none of our team who recorded the activity and behavior of each experimental animal minded being on duty for 24 hours—it gave them more opportunities to observe whales, sea birds, and Iñupiat on the hunt for seals in their skin kayaks.

With the animals, we learned that even when the sun was going in a circle overhead for 24 hours each day, the same regular wintertime sleep habits persisted during each summer of continuous light. They ignored the daylight and curled up with their tails over their heads and slept.

We decided to add wolverines, which are nocturnal, to our program. Someone had earlier given Max Brewer a baby wolverine and he had raised it, although it was kept in a cage and was not tame. When I mentioned to a Canadian Fish and Wildlife manager that I wanted to start a wolverine colony, he simply asked, "How many do you need?"

Our colony developed into one of the biggest wolverine collections in the world, despite that wolverines are difficult to cage, maintain and handle. The largest weasel on the globe, wolverines weigh about 40 lb, and like the rest of the weasel family, are quite fierce. These mammals are exceptionally strong for their size. We managed to study them, however, and found that, despite their high metabolism, they rigidly kept their 24-hr rhythm behavior pattern regardless of changes in daylight, just like other nocturnal animals.

## *Bear Hibernation*

We had primarily developed the implantable radio technique to study hibernation in black, grizzly and polar bears. We of course started with the smaller mammals to get experience for the bears that would come later. To study small mammals in hibernation, implanted transmitters were helpful but not essential. We had made many records "by wire" on small hibernating mammals, especially ground squirrels, hamsters and woodchucks. We would simply place an electrode on their backs to get to the heat of their skin—they were too dopey to care. Bears would not be so easy to handle!

At that time, international symposia on the topic of Mammalian Hibernation were held every four years. At these meetings, it was often mentioned that there was no information to settle the question of whether bears *hibernate* in winter or merely *sleep* for long periods. Up until about the mid-seventies, many biologists accepted the premise that bears do not experience true mammalian hibernation.

We were eager to use our Iowa implantable radio transmitter and surgical technique to prove whether or not bears hibernate. Over a 10 to 15 year period we accomplished our goal of studying bears in hibernation.

## *Immobilizing Drugs*

But to perform surgery on a large, irascible animal like a bear, we needed a tranquilizer that would have no ill effects on the animal. Drugs used earlier, particularly *succinylcholine chloride* had many disadvantages, one being that the dosage had to be precisely based on weight, which was difficult to estimate for large animals like bears, especially in the field. After several polar bears were accidentally killed by another biologist who was using succinylcholine, the Department of Fish and Game forbid its use in polar bears.

Then a drug came along that had no fatal dose. It was the new immobilizer drug *phencyclidine hydrochloride*, known as *Sernylan*. Because the drug had been put to illicit use on the streets (where it was called Angel Dust), clinical medicine had rejected its use in people. However, the drug's maker, Parke Davis, was interested in determining appropriate dosages for various animals, and supplied the drug without charge to our laboratory for many years. Thus we were the first to use the drug for this purpose, and we did work out the dosages for all Arctic mammals. The best thing about this drug was that it never acts to kill the animal; if one misjudged the dose of the drug or the weight of the mammal to be studied, the patient stayed unconscious longer but did not die.

Pt. Barrow, AK, Arctic Research Lab--

Inside the snowed-in quonset hut above, a bear hibernates in a nest of straw (as illustrated). A radio capsule inside the bear's abdomen transmitts its EKG. The implantable physiological capsule was invented by Profs. Folk (l) and Essler (r), shown below, and was used to obtain physiological data in many other species of mammal.

We began the process of implanting transmitters in bears by using an immobilizer gun with sernylan to dart the bear (we never used the word "shoot"). Immobilization only lasted about 20 minutes, during which we transported the bear, usually with Pete's help, to the surgical area where we used ether or chloroform as anesthesia. Then I proceeded to surgically implant the radio capsule in the abdominal cavity. Over many years, we implanted transmitters in all three species of adult bears, including a 1,000 pound polar bear. With bears that were too big to move, I performed the surgery right in the bear's cage.

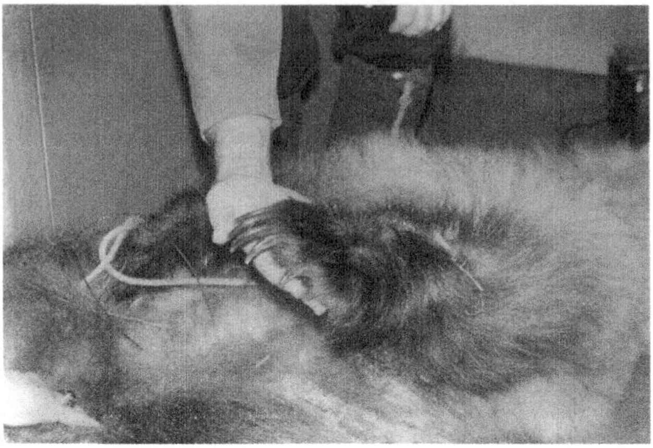

**Paws of a grizzly bear.**

The anesthesia was administered with a cone over the muzzle by my assistant—usually my wife, Mary. We often joked that Mary's job was to keep Ed safe, but the reality was that it was Mary's hands that were near the jaws of the bears!

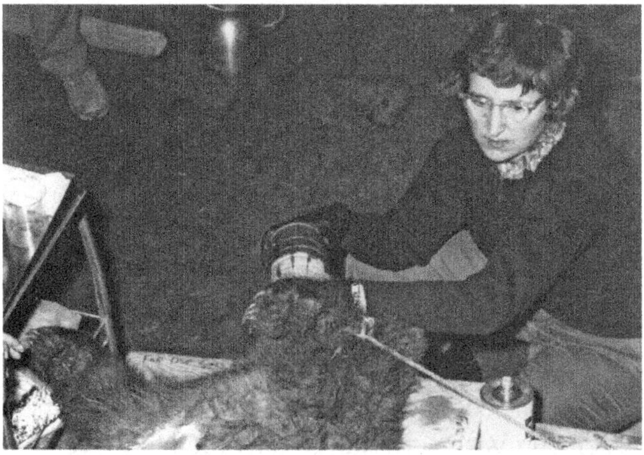

**Mary, bear anesthesiologist.**

I must explain that not everyone was happy to see Mary become a "bear anesthesiologist," and safety was not the issue. At one meeting back in Iowa, the dean complained that I had stolen the department's best medical illustrator and turned her into an anesthesiologist! His jabs were all in good fun, but the truth was that despite the good money she could make as a medical illustrator, Mary had gladly quit that job to become my hardworking, unpaid teammate.

### *Our Bears*

Our bear hibernation program began with the encouragement and help of the great naturalist and well-known zoologist, Marlin Perkins of the Chicago zoo (he also hosted *Wild Kingdom* on television from 1963–85). Perkins gave us an adult male black bear, which he shipped to NARL where Kenny Toovak and Pete Sovalik got him settled into his new environment.

Since adult bears could reach 1,000 lb, it was left to Kenny to build all the large, strong cages that would be needed for these huge beasts—no trivial task. Fortunately, Kenny was capable of just about anything. Mary and I later instrumented this black bear with a capsule to transmit body temperature and heart rate and placed it in a cold room where it made itself a den of straw. Within weeks it had a lowered body temperature and heart rate. For two years we successfully recorded the "classic" winter hibernation of this bear. It was the first definitive proof of "bear hibernation."

**Orphaned grizzly bear cubs.**

Later we used the same technique in two male grizzly bears given to us as cubs by Max Brewer. During the bears' second winter, we released them into a small building full of hay and recorded their activity, pulse and body temperature while they threw hay at each other all winter long, refusing to hibernate, probably because of the absence of their mother.

However, the next two winters they each made a nest in the hay and spent two months curled up in a classic bear-hibernation state. We recorded much-reduced heart rates in those grizzly bears, and in later studies found the same in polar and black bears. All the bears' heart rates dropped from 30–50 beats per minute (b/m) in summer sleep to 8–10 b/m in hibernation.

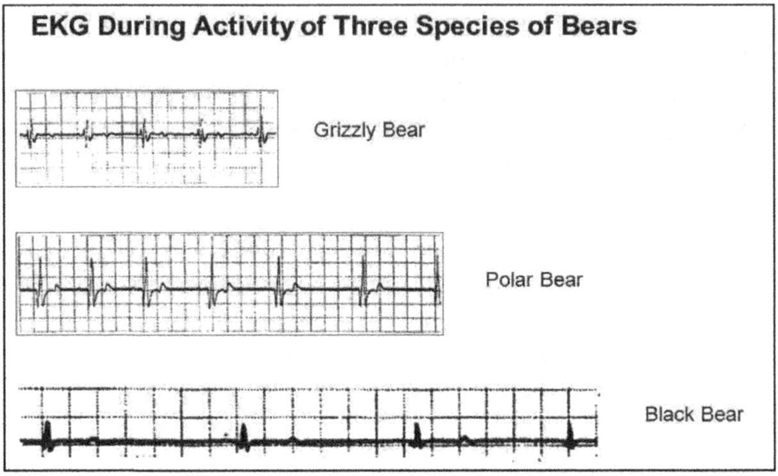


**The third EKG shows a black bear starting into hibernation.**

Thus, based upon an eight-year study of all three species of North American bears, our research showed that *the hibernation observed in bears is true hibernation,* even though *it differs from the hibernation observed in small hibernators*. One difference is that in hibernating bears, core temperatures do not drop as dramatically: the small hibernators will go as low as 5°C (41°F), while hibernating bears do not drop below 32°C (89.6°F).

**Female polar bears, like this one I photographed from my car, stay in hibernation longest of any mammal.**

The other main difference is that small hibernators arouse themselves every few days, warming, feeding and resuming bodily functions, while hibernating bears do not move very much during 4 to 7 months, nor do they eat, drink, urinate or defecate during that entire time.

Polar Bear EKG

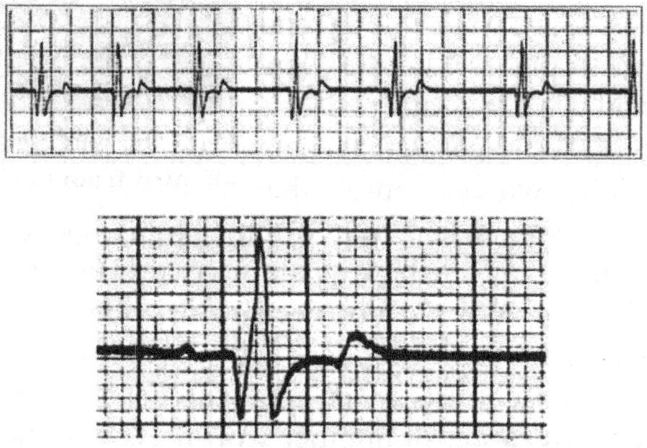

Taken by Radio Telemetry in Winter Den

We concluded that, in the evolutionary sense, the hibernation of bears is more perfected than that of small mammals, in part because they maintain their modified state for a much longer period than do the small hibernators (Folk, Larson, Folk 1976).

## *Wolves*

Pete Sovalik started our wolf program by bringing me six wolf pups he had captured near Anaktuvuk Pass in the Brooks Range. These Alaskan Tundra Wolves (*Canis lupus tundrarum)* are a subspecies of the gray wolf that inhabits North America. Their treeless tundra habitat requires different hunting and denning behavior than wolves living in a forest.

Over the years, the offspring from these six pups totaled seventeen litters. It takes a lot of work to provide wolves with raw meat and bones four days of the week and fish on two days, while they always had access to a feeder containing commercial dog pellets. Eventually we had too many wolves to maintain; we gave them away as fast as we could, because we didn't have enough cages to keep any more than 20 wolves at a time. The adult wolves weighed from 66.5 lb for a small female, to 109.5 lb for a large male.

**Kate Persons and the wolf she adopted from our colony.**

Fortunately, a lot of other people wanted wolves, so we were able to send many of them to places like the New York Zoo, and we also returned one pack to the wild. We released that pack of young-adult wolves into the Anaktuvuk Pass area where our original pups had come from. A graduate student, Bob Henshaw, spent several weeks there observing their reintroduction. We did not need to provide for them, as they quickly learned to hunt. Our biggest fear was that wild wolves would kill them, but because it was a vast area with many caribou migrating, there was plenty of room for another wolf pack, and the wild wolves ignored them.

One young wolf was adopted by our animal caretaker, Kate Persons, despite Max Brewer's opposition. Kate raised the pup to be a lovely mature animal, which, it was reported, slept in bed with her! An intrepid young woman, Kate had come with Mary and me from Iowa City to Barrow for a summer job in my laboratory, and never left. Even when I couldn't support her, other people did. A serious outdoorswoman and lover of animals, Kate raised and trained sled dogs, and competed four times in the famous Iditarod dog sled race from Anchorage to Nome. I had the privilege of being one of her dog-handlers for two of the races. Kate eventually moved to Nome and became a wildlife biologist for the Alaska Department of Fish and Game.

Of the wolves we kept, we allowed most of them to reach old age, sometimes continuing to record changes in their physiology as they got older. Their relationships with us were interesting: the dominant males and females would come to the front of the cage to be petted, showing to us the kind of fawning behavior lower-ranked wolves show to their Alphas; other individuals in the pack usually hung back.

## *Wolf Behavior: A Triangle among Wolves*

One year we conducted a study of courtship and mating behavior in one of our wolf packs. The results were surprising. Because our detailed observations chronicled rarely observed wolf behavior, I'm describing it here in more detail than I've given to other studies.

This pack of seven-year-old wolves was born at Barrow; the four males and one female were littermates with a well-organized hierarchy of dominance. The wolves were easy to identify because each was a different color: the *alpha* male (Mix) was grey and black, the *beta* male (Mr. B) was brown, the female (Charlotte) was solid black, one subordinate male was black marked white (Frosty), and the other subordinate male was grey (Rusty).

There was no question that Mix was the alpha male. Certain regular events occurred that showed the alpha status: (1) On many mornings all the other members of the pack would mob Mix, licking his snout in greeting. (2) Most days, Mix would lead the pack for ten minutes through all four quadrants of the enclosure as if going out on the hunt; his tail would be straight in the air in the usual alpha position, while the entire pack was subdued and followed him calmly. (3) Mix would end confrontations between other wolves. For example, if Frosty would not let Rusty eat from the feeder, Mix appeared and shouldered Frosty away. (4) Especially during times of excitement, Mix pinned Mr. B, the beta wolf, to the ground, holding him by his throat in a loose grip.

In wolf packs, the role of the beta male is unique, and not always matched by a powerful beta female. In both wild and captive packs, beta males are known to threaten other wolves including those of higher rank, yet they also associate closely with the alphas. In our pack, the alpha male, alpha female and beta male often slept together away from the subordinates, and were affectionate in their greetings with each other.

The wolves lived in a large outdoor enclosure divided into four equal sections with connecting doors that could be left opened or closed. Shelter was provided by three plywood houses.

We built a small observation building with an eye-slit through which we could continuously monitor each wolf and log its behavior without announcing our presence. At night, the full moon (or artificial lights calibrated at the same level) was sufficient for us to identify each of the five wolves. During the observation period, light snow fell occasionally, temperatures varied from 10°F to -20°F, and wind speeds ranged from 5 to 20 miles per hour.

**NARL wolf compound.**

Besides myself, the trained observers included Mary, Kate and two others. We took five-hour shifts to observe the pack around the clock, and kept detailed records of eating, sleeping, activity and interactions. Because wolves live in a world of smell, we included the wolves' reactions to the urine and feces of other members of the pack and the smelling and licking of genitalia. Our records totaled approximately 400 pages.

When the female came into heat she became very agitated, constantly paced down two sides of the quadrants, and sometimes hurled herself against the wall after pacing the whole length of the enclosure. She also stopped sleeping in the den with the two males.

Both the alpha and the beta males began to "court" the female. The alpha male, Mix, jumped at the female stiff legged, rushing and playing like a puppy. He would put his paws on her back without mounting, and Charlotte would lick his face. Mix's courtship lasted about one day, and Charlotte never appeared to reject him.

When the beta male, Mr. B, courted Charlotte in the same fashion, she returned his attention vigorously and aggressively, bouncing circles around him, constantly putting her paws on his back and even mounting him. During one such display, the alpha male growled and snarled at the beta male, but was ignored.

Mr. B began to join Charlotte at night in her corner; they usually slept with her head on his back. Soon Mr. B began to mount Charlotte, who eluded his grasp, but always licked his face. This was a critical time; Mix had been resting and watching, but from then on when he approached, the beta wolf growled and snapped his teeth. Mix respected these signs, but when the beta wolf was not close to Charlotte, he too would try to mount her. She never growled, but usually moved away.

Gradually the pair bond between Mr. B and Charlotte intensified with puppy games, wrestling on the ground, and rearing up with paws and chests together. If the alpha wolf approached during these games, Mr. B would rush at him growling with teeth bared. Mix responded with similar strong growls and snarls. At times the two males would leave Charlotte and pace through the area shoulder to shoulder, both growling and snarling.

All four males began to show an intense interest in the female's urine. When she deposited it on the snow, the males stood around it like spokes in a wheel, licking it up without any aggression. Mr. B would then go back to receiving the ardent advances of Charlotte. The alpha male finally accepted the situation, chasing away the subordinate males and seemingly standing guard while the alpha female and beta male were breeding.

At the end of the heat period, Mr. B appeared fatigued and slept away from Charlotte for several hours. Charlotte allowed Mix to mount her, but after licking his face she moved away. As soon as Mr. B reappeared Charlotte behaved like a puppy, playing with him, running circles around him, and putting her paws on his shoulders. The subordinate wolves now ignored Charlotte, who again had the free run of the enclosure without being aggressively solicited by any of the wolves.

Soon after the breeding period, Mix was again the welcomed companion, and both males' heads rested on Charlotte as they circled with tails wagging. During these interactions, the tail of the alpha wolf was vertical and straight up, while the tail of the beta—the breeder—was down and drooping. The three resumed sleeping in the den together, and a litter of pups was born 63 days later.

**Alpha male, Mix.**

What we had witnessed was the unusual event where it was the alpha *female* who decided which of the two highest ranking males would be the breeder. Breeding by the beta male, while not typical, has been observed, but only in captive packs. This may indicate aberrant behavior due to confinement, or it may highlight the difficulty of observing a wild pack over an entire breeding season. It would also be unusual for full siblings to breed in the free environment, as young adults typically disperse to other packs.

Wolves' mating systems include a great deal of variation. Mate preferences and social relationships are complex and can change from year to year, or be foiled by aggression. What was observed in our pack was an unusual example, but it shows that, no matter what, wolves will find a way to continue the species.

## *Fundus Photography and the Janes*

Dr. Ralph Janes from the Anatomy Department at the UI was a colleague and close friend. Mary and I enjoyed socializing with Ralph and his wife Louise, who later became my secretary and worked with me for many years.

It was Ralph who first told me about the immobilizer drug sernylan. I wanted to investigate whether changes in vessels or pigments in the back of the eye—the fundus—occurred with changes in seasons. Was there more pigment in the eye in winter? I asked Ralph to work with me on this project, and he obtained a hand-held fundus camera from the UI Dept. of Ophthalmology and learned how to operate it. He accompanied me to Barrow to photograph the fundus of several mammals. Sadly, Ralph died shortly after our return, and his slides were lost.

But another researcher, Ogden Frazier of Ophthalmology, took over the project. During several trips in 1966 and '67, we used sernylan to immobilize the animals and photographed the back of their eyes. In all, we took pictures of the fundus in wolves, all three species of bears, lynx, wolverines and foxes, as well as several smaller animals including porcupines, marmots and voles. The photographs, some of which are quite beautiful, found that there were no seasonal changes in the pigmentation in the fundi of any of these mammals.

## Fairbanks: Sabbatical Year

When I went through Fairbanks on my way to Barrow, I often saw my colleague and old friend, Larry Irving, who had founded the Institute of Arctic Biology at the University of Alaska Fairbanks.

Larry would always ask, "When are you moving here, Ed?" Of course I could not actually move there—I had tenure at the University of Iowa, and that was golden.

One advantage of being at a university, however, is the opportunity to spend a sabbatical year at another university or research facility. Sabbaticals are designed to give faculty members a break from teaching and administrative responsibilities in order to write books, undertake new research or study. Most of my American colleagues went to Europe or Australia. But, in 1964, when I was offered a sabbatical in Fairbanks, Mary and I immediately "moved" there so I could spend a year at the Arctic Aeromedical Laboratories (AAL). We drove up to Fairbanks from Iowa City.

The Air Force created a new position for my time there; I felt honored to receive the first Air Force Professorship ever given. It was a great position, and the sabbatical paid more than my job in Iowa. We spent a happy year in Fairbanks; I worked in the lab by day and on my new textbook at night.

Mary and I lived in an apartment that had an exciting view of Mt. McKinley, the highest mountain in North America (20,320 ft or 6,194 m). Originally called Denali, which means High One, the mountain is the heart of Denali National Park. It was only 120 miles from Fairbanks to Denali on one of the few good roads in Alaska; Mary and I were fortunate to be able to go there often.

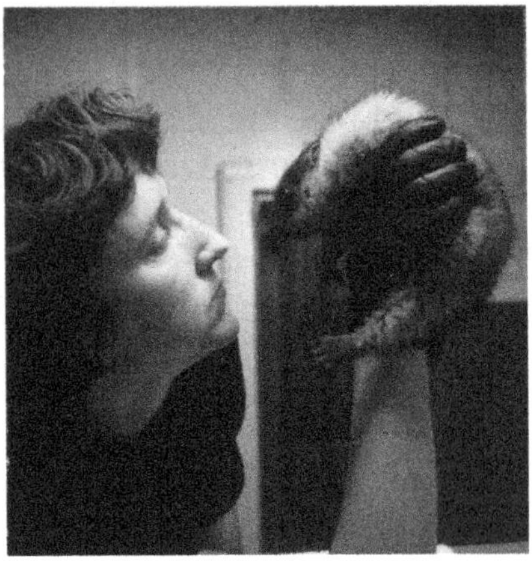

**Mary with a pet marmot.**

One of my projects was comparing weasels and rats on running wheels. Mary thought the weasels were cute, and she couldn't resist bringing one home from the lab. It was a New York weasel, a species that turns white in the winter; we fed it chicken hearts, played with it and watched it turn white.

During our study, one night a female rat ran 10,000 revolutions on a running wheel, or 21 miles—that great distance was because the animal was in *estrus*, or "heat," which in the free environment compels the female to travel fast and far to find a mate. This type of behavior in estrus is seen in some other species; indeed, many horseback riders have suffered the exploits of the mare in heat.

I once experienced the mare-in-heat phenomenon firsthand when I was riding a mare that was in estrus. We had begun to trot down a long sandy road when suddenly the mare seemed to go crazy, grabbing the bit and breaking into a fast gallop; she was out of control and resisted all my efforts to stop her. Ordinarily one tries to put a runaway horse in a circle to slow it down, but there wasn't room on the sides of the road and we were fast approaching a highway. I saw a long private drive and turned her into it hoping I could stop her in someone's yard. Instead, when she saw a garage built into the side of a hill, that mare ran straight up the earth berm to the top of the garage and out onto its roof, barely stopping at the edge. Stomping and pawing, she seemed ready to jump off the roof onto the paved driveway below. Taking advantage of the moment, I quickly slipped off and eventually led her back to her owner. She had been my favorite horse before that incident!

A conference in Skagway, a city in Alaska's panhandle southeast of Fairbanks gave us the opportunity to visit one of the first gold rush towns and to explore further east into the Canadian Yukon. We had heard it was a beautiful drive with good camping along the way, so we decided to drive the 710 miles between Fairbanks and Skagway. We were returning from our visit to the Yukon, and stopped to set up camp in a flat area near an abandoned gold mine. We were still in the truck when everything began to shake violently.

We looked at each other frowning. Earthquake?

The earth continued to shake for many minutes that evening of March 27, 1964, and thus Mary and I experienced the Great Alaskan Earthquake—it registered 9.2 on the Richter scale. The epicenter was south of us in Prince William Sound, and the shaking was felt all over Alaska, Canada and most of the US.

Fortunately for Mary and me, we were not in one of the areas affected by landslides and avalanches. But in Alaska, 115 people were killed, mostly by waves triggered by landslides in fjords and bays; another 16 people as far away as Oregon and California were killed by a tsunami wave. Alaska's major cities suffered structural damage, but the low population density kept the death toll and losses relatively low for such a strong quake—the second most powerful recorded at that time.

### *Hypothermia Studies*

While we were in Fairbanks, an opportunity arose to study hypothermia, a condition about which much is known and much remains to be discovered. In humans, for example, core body temperatures below 26°C (78.8° F) shut down the body's ability to shiver and spontaneously rewarm, often leading to ventricular fibrillation (VF) that stops the heart.

Small hibernating mammals, on the other hand, routinely allow their core body temperature to decrease to near 0°C (32° F). In one study of hedgehogs conducted in 1963, Swedish researcher Johansson found it was impossible to induce VF with hypothermia.

Studies into hypothermia have uncovered its usefulness as a lifesaver in certain circumstances. Moderate hypothermia of 32°–34° C (89.6° F–93.2° F) is now intentionally induced in the hospital to help patients survive cardiac bypass, as well as to protect the brain in those who have just suffered cardiac arrest. Our team's studies in hypothermia and hibernation in Alaska in the 1960s were among the many that contributed to this body of knowledge.

One of our hypothermia studies would be controversial today, and in fact would not be permitted. At that time, however, we did not have the same strict regulations we have today to protect human and animal subjects. Here is what happened.

I received a phone call from an Air Force veterinarian working at an airbase in Texas. He said, "We know you're keeping some bears up there in Alaska, so I felt duty bound to call you. We've been using black bears in an experiment, but now that it's finished, I have three leftover bears—will you take them?" He went on to say that the Air Force had been studying the effects of ejection cockpits using bears as a model for men because bears spend quite a bit of time standing upright on their hind legs like humans. At first I declined to take the bears because we didn't have room to include them in our hibernation studies—it takes a lot of space and resources to house animals that large.

However, when he added that these animals would be euthanized and their bodies taken to a "dump," I felt that as a bear biologist I should take them, even though we had only a temporary situation for them. I knew that I would be able to conduct the euthanasia humanely using hypothermia; thus this unhappy event would record physiological data that could prove useful in both human medicine and wildlife management.

The research question addressed was, "Can a large hibernator tolerate hypothermia at temperatures lower than other animals and people?" We followed all animal welfare regulations in effect at that time, just as we did in all of our studies involving animals.

We placed the bears under anesthesia before we began hypothermia. We took heart rates and other measurements as the bears' temperatures and finally their hearts stopped. The data we gathered was surprising. Although bears are hibernators, in hypothermia they were more similar to humans than to small hibernators, who are protected by their physiological condition from accidental or applied hypothermia. Therefore we learned that unlike small hibernators, both large hibernators and humans are susceptible to hypothermia-induced ventricular fibrillation.

## A Long Commute

When my sabbatical in Fairbanks ended, I returned to the University of Iowa, and soon published the textbook I had written during that year. *Introduction to Environmental Physiology* came out in1966, and I was later credited by the American Physiological Society with helping to define and establish this emerging field.

**Relaxing in Iowa with Alaskan furs.**

Based back in Iowa City, I resumed my routine of teaching and working in my campus laboratory, while also "commuting" to my lab at Point Barrow every chance I had. With the help of several exceedingly able graduate students who spent up to two years each in Alaska, I was able to not only continue, but also expand my research in the Arctic. This routine continued until 1977.

During these years I was productive, happy and deeply involved in my research into hibernation and circadian rhythms. I also somehow found the time to revise, update and publish the second edition of my *Textbook of Environmental Physiology* (1974).

## Adrenaline Rush on Bathurst Island

Sometimes my research in this unique landscape led to uncomfortable and even dangerous adventures. As I described earlier in this book, fear can cause a physiological response commonly called the *fight or flight syndrome*. The obvious pounding of the heart, shaking of the knees, and rapid breathing is caused in part by a high titer of adrenaline in the blood—hence the term, *adrenaline rush*. I have had this reaction only a few times in my long life: once while trapped in a narrow tunnel in a cave, once because of a near-disastrous auto crash, and once while looking for lemmings on Bathurst Island in the High Arctic.

Bathurst Island is in the Canadian Archipelago of the NW Territories (now known as Nunavut). With its latitude of 82°N, this uninhabited island is farther north than Barrow by at least 10°. To get to Bathurst from our laboratory on Point Barrow is a journey of almost 1500 miles by air or sea. I flew there in one of NARL's 2-engine transport planes.

The tundra of Bathurst Island has abundant low vegetation to support its wildlife, and it was the wildlife of Bathurst that gave me my reasons for wanting to go to this uninhabited High Arctic island. Bathurst is a unique place where several species have undergone distinctive adaptations. For example, the small white Peary caribou *(Rangifer tarandus pearyi)* on Bathurst has been selected by evolution to be all white for most of the year; in summer, its coat turns partly gray, with white legs and underparts. These small caribou are a distinctive subspecies found only on Bathurst and a few other Nunavut islands. Over the past several decades, the caribou at Bathurst have experienced a large decline in numbers due to especially harsh winters. (The Peary caribou is now listed as endangered under Canada's Species at Risk Act of 2003.)

Bathurst Island was also inhabited by very large arctic hares whose shortened ears are a physiological adaptation to the extreme cold. Arctic hares also occur on the mainland where they turn white in the winter and brown in the summer like the snowshoe rabbits in the south. However, on Bathurst, these hares remain white over the entire calendar year. This encouraged us to look for evidence of the retention of a white coat in the spring and summer by collared lemmings, a species of lemming found on Bathurst.

Thus, my primary reason for going to the High Arctic was to collect collared lemmings *(Dicrostonyx groenlandicus)* to establish a breeding colony back at Barrow, and to compare them with the brown lemmings *(Lemmus sibiricus)*in that region. Also, I wanted to answer a second question: was it possible that at Bathurst, collared lemmings might remain white all year instead of turning brown in the summer?

**Collared Lemming.**

Lemmings are small (about 6" long), brownish, chipmunk-like rodents that live on Arctic tundra and can survive the winter without hibernating. Lemmings are known for their dramatic population cycles—every four years or so the population simply explodes, then it takes a drop so precipitous that it approaches extinction. People don't know why this drastic cycle occurs; theories abound and entire books have been written on the subject!

Lemmings provide an important food source for most Arctic carnivores, including foxes, wolves, snowy owls, rough-legged hawks, weasels and even polar bears; thus they are said to "control the rhythm of Arctic life" with these cycles of population abundance and scarcity. Snowy owls, for example, depend on large numbers of lemmings to feed their young, and typically do not breed in years of lemming scarcity.

I learned quickly about high-population "lemming years" at Point Barrow because if a door was left open at home or in the laboratory, these little creatures would run right inside—they were so numerous they inundated all available space like a flood. In contrast, during the 4th year low, you could scarcely find a lemming on the tundra.

The same population dynamics occur with the collared lemmings, which may be found throughout the Arctic, but primarily in the High Arctic at Bathurst. Collared lemmings are typically brown or gray in the summer, with stripes down the back and on the sides of their heads. Commonly known as the "varying lemming," they turn all white in the winter (the only rodent to do so), and grow front claws for digging through heavy snow. Their brown cousins in Barrow remain brown or gray all year. Considering that both live above the Arctic Circle, why is this?

The pineal gland is known as an important regulator of biological clocks. It has an unusual "regulatory assignment" at Point Barrow with its 82-day cycles of continuous light and continuous darkness. Even when the pineal glands of 10 brown lemmings were removed, each of these wonderful, tough little animals not only survived but also retained its original body weight. I wanted to compare the biological clocks and pineal glands of the brown Barrow lemming with those of the varying lemming from Bathurst; I expected the latter would be larger because of the extraordinary task assigned to this particular endocrine gland at an environment even more extreme than Barrow.

My visit was also partly inspired by the outstanding zoologist and ornithologist, David Parmelee, who once showed me his laboratory collection of study skins of Bathurst Island birds, including baby sandpipers of various ages.

Another inspiration was the account published by David Gray, of Canada's Museum of Natural Sciences and a colleague of Parmelee. In the late 1960s, while Gray was on Bathurst Island to study musk oxen (*Ovibos moschatus wardi)*, he witnessed an amazing event: a lone white wolf was able to kill an adult and apparently healthy male musk ox. This alone was very unusual. Gray then went on to report that not only did the predator return to feed on the carcass all that summer, but also during the second summer, he observed the same wolf, this time with a female companion, make several forays back to feed upon the carcass (Gray 1970). This dramatic example supports the concept of "the economy of nature."

After a group of us had arrived by plane and set up our tents on this uninhabited island, I wandered over much of the terrain looking for lemmings. I soon found that it was the crash period in the four-year cycle for our white lemmings. The two lemmings I did see retreated into deep holes and were too far from camp for me to set traps. My live-traps near camp caught nothing.

I had one week until the plane's return, but it was not wasted because I had a rich experience photographing the wildflowers of the region, the musk oxen, the Peary caribou and above all, the arctic hares, as they stood on their hind legs to see farther over the landscape. There were many birds to photograph, including a red phalarope nest near my tent. The bird must have been a male, as the females delegate the nest building and brood raising to their mates! These small, brightly colored shorebirds actually prefer open waters where they fish by swimming in circles to create a vortex from which they scoop up plankton with their bills.

Most of the week I traveled the island with other people, but one day I went out by myself. Looking for lemmings, I walked up a frozen riverbed for half of the day. While I was eating lunch, a herd of musk ox wandered near, looked at me curiously, and then settled into a siesta. I continued my hike up the valley around them.

As I walked, I collected clumps of musk ox wool where the animals had rubbed it off against rocks and bushes. This silky underhair, called *qiviut* by the Inuit, is shed annually and is considered the finest natural fiber in the world, softer than cashmere or camel hair, and wonderfully insulating. I was soon able to fill my pocket.

**Herd of musk oxen. (Source: US Fish and Wildlife Service.)**

As I came around a sloping rock, I came face-to-face with a single, angry-looking male musk ox; he may have been ostracized from the herd I saw earlier. I clambered onto the rock because this 900-lb animal was showing the three signs of aggression: pawing, shaking his head, and snorting. But after quite a display, he decided not to charge; he just turned and went off in the other direction. I came down off the rock, excited and wide awake, but not having experienced an adrenaline rush.

Engrossed with photography and still hoping to see a few lemmings, I wandered farther up the valley, coming out on the opposite side of the frozen riverbed. I realized I had gone quite a few miles and it had become fairly late in the day, so I must return to camp. I also realized that I had walked on both sides of the river valley, and to retrace my steps would take far too long. In this area, the valley floor was a wide expanse of snow and ice.

At about this point, I found a musk ox trail back across the icy valley. It was distinctly marked by their hooves in the snow, and I decided that if it was strong enough for the musk ox, I could use their crossing. But about two-thirds of the way across a smooth area of snow and ice, my left foot and lower leg went through the crust to the top of my knee-length rubber boot.

As I pushed to withdraw it, the other boot went through the ice in the same way! I knew I was in serious trouble when I heard the sound of rushing water below me, and realized I was on an ice shelf over the submerged river. My rubber boots were fixed in place in the icy crust, and I raised my legs out of them.

As I recognized my predicament, the adrenaline rush began. At any moment, a piece of the ice roof might fall through with me on it, and I would be in icy water, flowing rapidly away within an ice cavern. I presumed my companions would send out a search party and might find a hole in the ice, but they would certainly never find me.

My first thought was, "Get to shore"! I did not dare take steps, but crawled on hands and knees to reduce surface contact. As I sat there on a rock contemplating a three- or four-hour hike back to camp in stocking feet, the adrenaline rush continued. I had to retrieve the boots.

Returning to hands and knees, I crawled back over the ice to the two holes and gingerly extricated both rubber boots, wondering whether this would weaken that shelf of ice so it would surely fall. The sound of rushing water seemed louder than before. Inching my way to shore dragging the boots, I sat down again on the rock and allowed the adrenaline rush to subside.

Boots on, I headed back toward camp. As I hurried, I thought of my project assistant back at Point Barrow, Bruce Ehrenhaft, who was my only contact from Iowa at the Laboratory. Bruce was a very conscientious graduate of Lawrence College, who had traveled 4,000 miles to his new job at Barrow carrying his violin as hand luggage. I could not help but wonder how that sensitive young man would have selected the words to tell my wife and daughter that I had disappeared into a hidden, ice-covered river flowing down a mountain.

I returned to camp safely. After I told my story, a veteran of past visits to Bathurst Island told me I was lucky.

"Oh yes, I know that crossing over the river," he responded. "We found a musk ox caught there with all four legs in holes where he had broken through the ice ceiling over the river. He would have starved to death or broken through into the water. But we called the Icebreaker at Resolute. They came with a helicopter, put ropes around the beast, and moved him to shore." I could not imagine how the crew placed ropes around the animal without falling through into the river.

I've continued to have many exciting adventures, and I'm happy to say that being trapped in the ice on Bathurst Island was the last time I experienced the intense fear of an adrenaline rush.

## Continuing Arctic Research

I closed my laboratory at Point Barrow in 1977, although NARL remained a strong force for Arctic research. However, in 1984, the Office of Naval Research's 30 years of support for NARL ended. The building at Point Barrow was transferred to the Ukpeagvik Iñupiat Corporation (UIC) and eventually became the *Arctic Sivunmun Iḷisaġvik College*.

This loss of an important research center at Point Barrow was a temporary setback to Arctic research until another facility there was converted into laboratories and housing for visiting scientists who want to study topics like Arctic biology and geology.

Now, in 2012, there are more opportunities than ever for those who wish to do Arctic research. Just a few of these resources include the following:

- Institute of Arctic Biology (founded by Lawrence Irving in 1962), University of Alaska Fairbanks
- Office of Polar Programs, National Science Foundation
- Arctic Research Program, National Oceanographic and Atmospheric Administration
- Most branches of the US Armed Forces
- Byrd Polar Research Center, The Ohio State University

There are many more, both in the US and internationally. And, one should not forget that in Barrow, Kenny Toovak and the Barrow Arctic Science Consortium will do whatever they can to facilitate research in the Arctic.

# *Epilogue: Horizons Near and Far*
**1978–2012**

## Continuing Work and Travels

I HAVE BEEN at the University of Iowa since 1954, and during all this time was able to continue my travels around the globe to conduct research, teach and attend meetings. I consider the years when I split my time between Iowa and the Arctic to be among the best. I also spent a year in the Netherlands on a Fulbright Fellowship. Mary and I didn't need to buy a car that year—we biked everywhere and used trains if it was too far to bike.

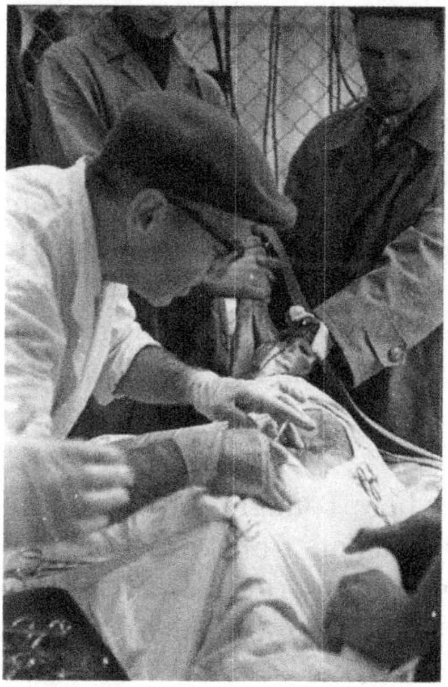

**Demonstrating surgical technique in Poland.**

Mary accompanied me on two trips to Poland where I enjoyed spending several months giving lectures at Jagiellonian University in Krakow. Part of my purpose there was to teach Polish scientists my technique for implanting small radio transmitters like the ones I had used in the Arctic to study the physiology of unrestrained animals.

A highlight of one of these trips to Poland was meeting with the highly respected wildlife biologists at Bialowieza Forest, a primordial woodland that centuries ago was the favorite hunting ground of Polish Kings and Russian Tsars. I was delighted by the large number of animal species in the forest—1500—and thrilled to see Poland's famous *zubr*, the European Bison that nearly became extinct until saved by a special breeding program at Bialowieza. I was also enchanted to see a rare herd of small forest horses—descendants of the extinct wild steppe horses or *tarpans*, the prehistoric wild horse considered to be the direct ancestor of all living breeds. Bialowieza is now a UNESCO World Biosphere Reserve.

Both times I went to Poland, I always felt like I'd come home. Why? I'm not of Polish descent, and thus I think it was because of the extreme friendliness of the Polish people everywhere we went.

We also spent some time in Iceland, where Mary and I had the opportunity to ride the famous pony-size Icelandic horses to experience their unique, smooth and very fast gait called the *tolt.* We were prepared to enjoy these small, pretty horses, but not only did we never experience the famed gait, we also found these horses to be headstrong and uncontrollable when turned towards their barn. Experienced riders though we were, neither of us could stop these thick-necked little beasts who ran away with us through ditches and other rough terrain to get back to their corrals and their hay! We decided not to import any Icelandic horses to our farm in the States.

I was also able to make some comparisons between the Arctic areas of Iceland and Alaska. In Iceland, we inspected the reindeer, which are domesticated smaller cousins of Alaska's wild caribou. And I also witnessed the Icelandic method of beaching and "processing" a harpooned whale, which was quite different from the same event undertaken by the Iñupiat in Barrow. For example, in Iceland the whale was pulled up ashore by several large tractors, whereas in Barrow it had been the combined muscle power of everyone in the village. And while in Barrow, all the villagers joined in cutting up the whale by hand, eating raw blubber as they did so, in Iceland, "professional" butchers were called in to perform the task.

I also spent time in Puerto Rico, where I taught researchers to implant transmitters to record physiological data from wild monkeys. Over these years, I never turned down the opportunities offered by environmental physiology to travel the world and do fieldwork on the far horizons.

## Prairie Knoll

In the late sixties, Mary and I purchased a small abandoned farmstead near Iowa City where we could keep horses, feed birds and have a fireplace. The old farmhouse required the addition of plumbing and electricity, and we lived there happily for years making few other improvements.

There was a big old red barn with a steeply sloping roof, and about 20 acres of rolling pasture. Behind the barn stood ten acres of woods, and five giant old European elms covered the wide lawn.

**The old farmhouse at Prairie Knoll.**

It was a birders' paradise; the farm and surrounding land attracted an array of songbirds, as well as a few raptors: a great horned owl, screech owls, barred owls, and a rare long-eared owl. The only nesting American kestrels in Iowa (at the time) chose our farm for several years in a row. Two of the most colorful songbirds, the Baltimore oriole and rose-breasted grosbeak, nested in our large elms every year. In the fall, after a night of high winds, I often found deposited on the lawn the oriole's delicately woven hanging nest. I kept records of migratory and resident birds, and by late spring, I could count 37 species, and hear them all singing simultaneously.

Mary and I rode our horses down the gravel road that ran from our farm to the Iowa River. We sometimes packed our supper and rode through the woods to a perfect spot on the hillside where we could picnic and enjoy expansive views over a neighbor's thousand acres. We had Mary's mare, Contessa, and quickly acquired other riding horses as well as two ponies for the grandchildren. Later we bred Contessa to a champion quarter horse stallion, and 11 months later welcomed an all-red filly. What a beautiful color she was! Copper Penny grew into a lovely riding horse.

Early one morning I awoke to the sound of hoof beats and saw, across the pasture, our young grandson Joshua (Victoria's son), astride a pony running headlong for the barn. Once the pony stopped, the boy slid off the animal's bare back, and still holding the bridle turned and led the reluctant pony back across the field. I was then surprised to see little Josh climb the barbed-wire fence, leap onto the pony's back, and hang on as the little beast galloped back to the barn! This amusing scenario was repeated several times, and I was pleased to see young Josh's spirit, spunk and resourcefulness.

**Little Josh and his sister Emily horseback with me on the farm.**

Another summer Kit's teenaged son, Chris Jon, needed a summer job, and I hired him to help me with several carpentry projects. He stayed most of the summer. I always enjoyed having grandchildren come to the farm, especially as I know how good it is for children to spend time in nature.

**Engaged in a project at the farm with Kit's children, Chris Jon and Lisa.**

Much to my delight, I found that the farm was inhabited by bull snakes. Although many people consider them aggressive, these were so mild-mannered that I could pick one up in the field (it might be five feet long) and it would allow itself to be handled before being released again. This always reminded me of my undergraduate days when I had earned money in Boston lecturing on snakes; my main illustration had been a live bull snake from Iowa.

From the farm, I could be at work on campus in about 30 minutes, thus we spent most summers there. We went to the farm year-round, but spent more time in our house in town in the winter.

## Mandatory Retirement—Not Really!

The University had a policy of obligatory retirement at age 70, as did many universities in 1984. However, the head of the physiology department invited me to stay on in my office and laboratory as long as I wanted—without salary, of course. I had several projects underway, so I took him up on his offer. It became obvious to me that I worked for the love of the science, though this is not to say that a paycheck would not have been welcome!

I continued to work full time and maintained a robust research program, although it meant I had to self-finance, or bring in money via grants from the National Science Foundation and other agencies. I also occasionally received small grants from our department, as well as some travel funds. I continued to publish and present research each year at professional meetings.

The same thing happened to my friend and colleague, James Van Allen, who became famous for his discovery of the Radiation Belts surrounding the Earth along its magnetic lines of force. I met "Van" when he came to one of my lectures on biological clocks before I went to Barrow. He had become interested in trying to determine whether subtle environmental factors such as the daily shower of radiation from space could be a clue that kept the circadian rhythm of plants and animals on a 24-hour basis. I attended many of his lectures on magnetospheric physics, a new field of science that came about through his discoveries. I met up with him twice in the Arctic, where he was sending balloon-launched rockets into the upper atmosphere to measure radiation levels from cosmic rays. When he came to Barrow, I was his "tour guide." Later he sponsored my membership in the Cosmos Club in Washington, DC. When Van retired, he maintained his office and laboratory in the Department of Physics and Astronomy, making good use of that space to the end of his days, publishing as many papers as ever.

The mandatory retirement policy changed within a few years, but it was never made retroactive for those of us who continued to work full time.

## Great Losses

Mary had always had a weak heart, despite which she rode horses, accompanied me to the Arctic, and spent two years living abroad. Her health did not allow her to bear children, however, and when it began to decline in 1982, we moved from our house on North Lynn Street to an apartment in the Oaknoll Retirement residence. There, we could still live independently, but we foresaw that, especially when I was out of town, Mary might need help, including medical care, which would be available from such a place.

We kept our farm, Prairie Knoll, and for many years, we continued our habit of staying in town most of the winter, and spending weekends and summers at the farm. Mary continued to be active; when she died suddenly in 1993, I was in the lab, and she had just finished carrying groceries up four flights of stairs. She was only 64.

Losing my beloved wife and companion was devastating. My comforts were my daughter, son, grandchildren, many good friends, and of course my work. Eventually, with the help of my former student, Professor Bud Riedesel, as well as writer/editor Diana Thrift, I was able to update, revise and publish the third edition of my textbook, *Principles of Integrative Environmental Physiology* (1998). Many of Mary's hand-drawn illustrations continue to grace its pages.

Eight years after Mary's untimely death, another tragedy struck. My dear son Christopher died unexpectedly; he was also in his mid-sixties. He had a long career at the Library of Congress in Washington, DC as an expert in Russian Literature, and was also a translator for the Navy. Yet, while Kit accomplished much in his professional life, I will always remember his love of, knowledge of, and even, one could say, rapport with turtles. From the age of four, he would go tramping through wetlands with sneakers and binoculars to observe the ways of turtles. As an adult, he had described two new species of turtle not known to be in Iowa. Kit and I had shared a bond through our love of nature, and his death was a blow. He left behind three lovely children who are now adults: Lisa, Julia and Chris Jon.

## The Antarctic

I went to the Antarctic about seven times, financing those trips by giving lectures about Antarctic biology on tour ships heading to that southernmost continent. On the first of these trips, I noticed an unusual occurrence in the flock of one million penguins on Paulet Island—there were no eggs in their nests. Why was this? That observation became an informal research project with the famous Antarctic ornithologist, Professor David F. Parmelee. After several years where egg-laying returned to normal, Parmelee and I theorized that ice storms can interfere with reproduction and nesting behavior, which appears to have been what caused the thousands of empty nests the first year I saw the penguins.

I enjoyed these trips to Antarctica because it allowed me to compare and contrast both Polar Regions, the Arctic and Antarctic. The geography is very different; for example, the North Pole is situated in the middle of a shifting ice pack, whereas the South Pole is in the center of a huge continent covered by ice two miles thick. That ice cap, consisting of layers perhaps 20,000 years old, is capable of oozing out ice shelves which, when they break, results in icebergs that can be 90 miles long, 25 miles wide, and several miles thick.

**Adele penguins on Paulet Island, Antarctica.**

The Arctic and Antarctic both experience months of continuous darkness followed by months of continuous light. However, when it is summer solstice at the Arctic Circle and the sun is above the horizon for 24 hours, the reciprocal is occurring at the Antarctic Circle, where winter solstice brings total darkness for 24 hours.

Another big difference between Antarctica and the Arctic is that while Antarctic seas are rich with marine life (whales, seals, penguins and other sea birds), there are no land mammals or indigenous people living on the Antarctic ice. What a contrast this is with the Arctic, where marine and land mammals both flourish, along with humans, who have inhabited the region for thousands of years.

Traveling to Antarctica, one has to sail 'round Cape Horn at the southernmost tip of South America, and cross the worst, most dangerous stretch of sea in the world. This is the Drake Passage, a strait between Cape Horn and Antarctica that is notorious for its fierce winds, rogue waves and icebergs.

Once, aboard the *Society Explorer,* as we approached Cape Horn, we heard from the Chilean Naval Station that as far as they knew, there would be clear weather. We were hoping to disembark and climb the hill to the top of the plateau to view Cape Horn. However, we missed our walk and the views because a violent storm came up from the west and the ship was soon rocking and pitching violently. The winds were 70 knots or higher, and the captain said the waves were 40 feet high. At first, the west wind was driving us toward some jagged rocks, but the valiant motors of this little ship persevered and we finally headed off and got back on course, still pitching and rolling. Surprisingly, some of the 500 passengers and crew had not realized the danger.

The next day, I talked with a passenger who worked for a British insurance company.

"That was a pretty close call last night," he observed.

I answered, "Well, I thought the motors handled that very well."

"That was not the problem," he replied. "The inclination meter on the ship read 15°, and when that happens, there is danger that the engines will come loose and break a hole through the hull."

He went on, "I work for Lloyd's of London and we lost two ships last year through just that kind of accident."

I stored this as an interesting comment, and glad to escape that fate, went on with my business.

Two years later, I was accompanied by my daughter, Victoria, on a ship of similar size, the *World Discoverer*. As our ship approached Cape Horn, exactly the same events occurred: the Naval Station did not know a storm was brewing; we could not climb to the plateau to view Cape Horn; and once again, the inclination meter read 15° and we were pitching and rolling violently. We were being pushed in the direction of the rugged coast, where the jagged rocks, with their white waves breaking high in the air, reminded us of snarling wild beasts.

The whole ship was medicated for sea sickness and bunked down in their cabins—including the Captain, a Dane who had made the trip over 90 times! Victoria and I, still healthy, were famished, and walked into the dining room. As we sat we noticed the tablecloths had been wetted down so dishes wouldn't slide off. We were the only passengers in the dining room, and no one came to wait on us for 15 minutes. Finally a waiter staggered in, his face a pale green, saying there was no food other than pre-made sandwiches.

Later, we were both standing outside the dining room holding onto metal pillars. At this point, an event occurred which made me think of the game called Tiddlywinks. An armchair was thrown from the far side of the lobby, across and through the air, to land hard on the pillar above my hands. Victoria and I looked at each other, eyebrows raised as high as they could be.

**Victoria at the rail of our ship in calm weather.**

After that exciting crossing, we enjoyed our visit to Antarctica. We saw four species of small penguins as well as the three-foot tall King Penguins who carry and incubate their eggs on their feet. We saw the wandering albatross gliding over the ocean on wings that span eleven feet. From the ship we saw right whales, killer whales and leopard seals—those beautiful and fierce top predators of the Antarctic (they share this title with killer whales).

In descriptions of leopard seals, the words "snakelike," "reptilian" and "evil" are used along with "agile," "inquisitive" and even "delicate." They have been known to stalk people and attack boats; there are several documented accounts of attacks on humans. Thomas Orde-Lees, a member of Ernest Shackleton's crew, was pursued by a determined leopard seal from both above and below several ice flows; the attack only ended when another crewmember shot the animal, which they estimated weighed more than a thousand pounds. Recently, a British marine biologist was drowned when she was grabbed and pulled underwater by a leopard seal. While scientists debate whether leopard seals' behavior towards humans is truly predatory or simply curiosity gone awry, it is clear that these animals are best viewed from a distance.

When Victoria and I were on the way home from Antarctica, we again experienced the wrath of the Drake Passage. This time the ship's engines stopped, and the motionless propellers were useless as we pitched up and down in icy waters. Our friend, the ship's engineer, sent a message to be prepared to abandon ship as we were drifting dangerously and rapidly toward the rocks on Cape Horn. Fortunately, he got the engines running just in time to get the ship headed safely into the wind.

## Bess

Several years after Mary's death, my good friends, Ian and Jan Smith, asked this lonely widower to accompany them to a dance because a nice lady, Bess Kueny, needed a dance partner. I too decided she was nice—very nice; Bess and I were married the spring of 1998, and I inherited her large family of seven children, including her wonderful daughter, Rosemary and brilliant son, Chris, both of whom have become very dear to me.

Before I married Bess, I often worked long days and well into the evening. But Bess made me agree to come home at 5:00 each night. Despite these "short" workdays, I was still able to accomplish a lot.

Soon after we were married, Bess and I circumnavigated the British Isles on an Audubon Tour. We went as far west as the Outer Hebrides, to Staffa and St. Kilda, where one can find Viking artifacts and descendants of Viking sheep. There the culture of some human inhabitants was to live on sea birds, collecting their eggs by roping down the cliffs. On the trip, we took pictures of puffins and gannets at our feet, and especially liked hearing nightingales sing.

That was when we met our "most interesting birder," Phoebe Snetsinger. Earlier Phoebe had been diagnosed with a terminal illness, and believing the end was near, began traveling with a passion to add to her "life list" of birds; 17 years later Phoebe was still going strong. On this trip she was hoping to add a particular bird—the corncrake. Bess and I felt privileged to be standing beside her in the garden of a beautiful monastery on one of the Orkney Islands, when she heard and saw her first corncrake.

Phoebe proved that birding is not for the faint of heart: she nearly died at the hands of five men in New Guinea in 1986, but was not deterred from her quest. Phoebe later died with her binoculars in her hand, the victim of a bus crash in Madagascar, where she was looking for an Appert's greenbul. She held the world record for bird sightings with more than 8,400 species; her book, *Birding on Borrowed Time,* was published posthumously in 2003.

**Bess Kueny Folk. Victoria and I are standing behind. I certainly was happy there with my two best "girls."**

Bess had loved to dance ever since she was a little girl, and we went dancing at least once a week. She had not had the chance to travel earlier in her life, so we made up for that by going to England, Alaska (including Pt. Barrow), the upper Amazon and the American West, where on one trip, we followed in the footsteps of Lewis and Clark. The two of us especially liked going to Washington DC, and we often stayed in the Cosmos Club. This allowed us to easily attend retrospective exhibits at the great art galleries in DC.

Bess lived her last several years battling *light-chain disease*, a rare blood cell disorder that required periodic regimens of chemotherapy. Though her life became filled with tiring treatments and recoveries, she remained optimistic and active. She volunteered at the University Hospital reading to sick children, playing games with older ones, and delivering mail and flowers to patients throughout the hospital. She attended writing groups, book discussions and dinner clubs where she was valued for her intelligence and perception. Always young at heart, Bess continued to dance until her death in 2009. Bess was a joy; it was shattering to lose her.

## Retirement...

Twenty-eight years after my official mandatory retirement, sixty years after coming to the University of Iowa, I still have an office and small laboratory in the Department of Molecular Physiology and Biophysics, where I enjoy the stimulating environment created by the talented scientists in this department. I continued to do research and publish, and several years ago, I was fortunate to gain several new colleagues and friends.

This came about because Eric Dickson—MD, researcher and then-head of the Department of Emergency Medicine at the UI Hospitals and Clinics—was interested in the cardioprotective effects of hibernation. Expanding on earlier findings of an opioid-like substance in the plasma of hibernators, Eric's ultimate goal was to find a similar substance that will protect humans against damage to the heart and brain during cardiac arrest and other emergencies, as well as during surgery. Eric invited me to participate in those investigations, and began to provide some support for my laboratory.

Dickson sent me two of his brightest clinical residents who he felt would benefit from the experience of working on a research project. It was immensely rewarding to work with these two young doctors, Eric Nilles and Christopher Buresh. While both showed great talent for research, they ultimately returned to their primary passion, which is to provide medical care to people in areas where none is available. Eric Nilles joined Doctors without Borders and worked for years in several war-torn countries in Africa, and Chris Buresh expanded his work in Haiti.

When we met Chris Buresh, he was already a veteran of several medical missions to Haiti, taking time off at least twice a year to volunteer his time and expertise. Chris is a man who cares deeply about people and it shows. He established friendships with many people in the town of Léogâne, about 20 miles west of Port au Prince, where he and his group conducted clinics. In January 2010, Chris had just returned to the US from Haiti when the devastating 7.0 earthquake struck, and Léogâne was at the quake's epicenter. Chris returned immediately to treat patients and help set up a field hospital. After at least a dozen more trips, Chris co-founded the Community Health Initiative to carry on this important work in partnership with the Haitian people. (For more information, please visit www.chihaiti.com)

**Eskimo-carved ivory polar bear, part of collection I donated to Smithsonian.**

I am now in my 98th year, and my research is winding down. I have begun to pare down the papers and books in my office and will soon turn the space over to one of our department's younger researchers.

In April 2012, I was honored by the American Physiological Society at their 125th Anniversary meeting in San Diego, CA, where I was given an award for an historical article about the Harvard Fatigue Laboratory during WWII. But I was most honored and proud to have my daughter, Victoria, accompany me.

I am fortunate to have been able to achieve my goal of *combining an outdoor life with research.* As I think back to the origins of my career, my mind focuses on some of the individuals who I emulated. The list begins with Larry Shields, my Biology teacher at Phillips Andover Academy, and goes on to include John Welsh, Donald Griffin, Harwood S. Belding, Bruce Dill and Larry Irving. These and many others contributed in ways both big and small to inspire my career in science and academia. I take this moment to honor them all. Without their examples, enthusiasm and guidance, I would not have had the many wonderful opportunities and adventures recorded in this book.

I know that the years ahead will continue to be full and busy, with nine great-grandchildren to celebrate, books to be read, birds to be watched, and more challenges to be met.

*End*

# *Acknowledgements*

This book would not have been possible without the early influence of Larry Shields, the Biology teacher at Phillips Andover who constructed the school's conservatory where I spent many happy years. By making me curator of this extensive conservatory, he kept me immersed in the world of field biology as a teenager.

My advisor and mentor at Harvard, John Welsh, not only kept me busy and inspired throughout my undergraduate and graduate years, but also arranged for me to spend a summer doing field work in the Caribbean. He set a high standard as a teacher, and I owe him much.

Two organizations deserve special recognition for supporting my work. First, my heartfelt thanks go to the University of Iowa, College of Medicine, and above all to Kevin Campbell and the Department of Molecular Physiology and Biophysics, for decades of unwavering support. Also, my sincerest appreciation goes to Eric Dickson, Christopher Buresh, and the UIHC Department of Emergency Medicine for many years of successful collaboration and funding.

Inspiration and support came from my family and friends who understood why I chose field biology rather than a career in a more quiet or lucrative arena. They deserve kudos from me.

And finally, I thank my editor, colleague and friend, Diana Thrift, without whose impetus, resolve and research skills I could not have completed this book.

# *References, Further Reading*
## Organized by Chapter


### *Prologue*

Fawcett, Don W. 1942. The amedullary bones of the Florida manatee (Trichechus latirostris). American Journal of Anatomy. Volume 71 issue 2, pp.271-309

Pearson O. 1942. On the cause and nature of a poisonous action produced by the bite of a shrew (Blarina Brevicauda). J Mammal 23:159-66.

### *Ch. 1 (Boyhood)*

Griffin DR. 1985. Reflections of an Experimental Naturalist. In: *Leaders in the Study of Animal Behavior: Autobiographical Perspectives*, p 120–142ed. Dewsbury DA. Cranbury NJ: Associated University Presses.

### *Ch. 2 (Undergraduate)*

Welsh JH. 1953. Excitation of the heart of Venus mercenaria, Arch. exp. Path. Pharmakol. 219, pp. 23–29.

Folk GE and Mora F. 2003. Serotonin and the Last Great Bioassay. Journal of Biological Sciences 3 (10):952-960.

Griffin DR. 1946. Mystery Mammals of the Twilight. National Geographic, July:117-134.

### *Ch. 3 (Salisbury School)*

Cannon WB. 1915. *Bodily Changes in Pain, Hunger, Fear and Rage.* New York: Appleton-Century.

Folk GE Jr. 1939. Easter on Wassaw Island. The Oriole, Quart. J. Georgia Ornithol IV(3):21-23.

### *Ch. 4 (Graduate School)*

Connell WA. 1981. Two New Neartic Epuraea (Coleoptera: Nitidulidae). Coleopterists Bull, 35(2):229-233.

Fawcett DW. 1997. Bloom and Fawcett: a textbook of histology. 12th Sub edition. London: A Hodder Arnold Publication.

Folk GE Jr. 1940. Shift of population among hibernating bats. J. Mammal. 21:306-315.

Folk GE Jr. 1940. The longevity of sperm in the female bat. Anat. Rec. 76: 103 109.

## *Ch. 5 (West Indies)*

Sanderson IT. 1939. Caribbean Treasure. The Viking Press, ISBN 0-670-20479-X. (Pyramid Books, 1965, 1966)

## *Ch. 6 (Harvard Fatigue Laboratory)*

Belding HS, Russell HD, Darling RC, Folk GE. 1947a. Thermal responses and efficiency of sweating when men are exposed to extreme cold. Am J Physiol 149: 204–222.

Belding HS, Russell HD, Darling RC, Folk GE. 1947b. Analysis of factors concerned in maintaining energy balance for dressed men in extreme cold. Am J Physiol 149: 223–239.

Byrd Polar Research Center Archival Program, The Ohio State University Libraries. 2010. http://library.osu.edu/sites/archives/polar/wilkins/wilkins.php (accessed 2/12/2010).

Consolazio FC and Forbes WH. 1946. The Effects of a High Fat Diet in a Temperate Environment. J Nutr 32(2): 195–211.

Dill DB. 1966. The Harvard Fatigue Laboratory: Its Development, Contributions, and Demise. Circ Res Vols 20 and 21 (Supple 1): I-161–67.

Dunar AJ, McBride D. 1993. Building Hoover Dam: An Oral History of the Great Depression. Reno: University of Nevada Press, p 45–47.

Gagge AP, Burton AC, Bazett HC. 1941. A practical system of units for the description of the heat exchange of man with his environment. *Science.* Nov 7; 94(2445):428-430.

Giesbrecht and Wilkerson. 2006. Hypothermia Frostbite And Other Cold Injuries: Prevention, Recognition, Rescue, and Treatment, 2nd Edition. Seattle: The Mountaineers Books. 141 p.

Griffin DE, Folk GE. 1944. Harvard Fatigue Laboratory Technical Report No. 30: Design objectives for electrically heated gloves. July.

Horvath SM, Folk GE, Craig FN, Fleishmann WF. 1948. Survival time of various warm-blooded animals in extreme cold. Science 107:171–172.

Horvath SM, Horvath EC. 1973. Harvard Fatigue Laboratory: Its History and Contributions. New Jersey: Prentice-Hall. 182 p.

Leigh OC. 1946. A Report on Trench Foot and Cold Injuries in the European Theater of Operations, 1944–1945. Annals of Surgery, Vol 124( 2):301–313.

Siple PA, Bazett HC, Pring RI. 1944. The use of impermeable barriers to prevent the destruction of dry insulation in footgear as based on field trials at Torbay, Newfoundland (cold-wet). (Fatigue Laboratory In-House Report) March 27–31.

Suleiman, SR. 2006. Crisis of memory and the Second World War. Cambridge MA: Harvard University Press.

Tipton CM. 1998. Contemporary Exercise Physiology: Fifty years after the closure of Harvard Fatigue Laboratory. Exerc Spotr Sci Rev 26:315–339.

## *Ch. 7 (Bowdoin)*

Chatfield PO, Lyman CP, and Irving L. 1953. Physiological adaptation to cold of peripheral nerve in the leg of the herring gull *(Larus argentatus).* Am J Physiol. Mar; 172(3):639–44.

Folk, G.E., Jr. Body temperature of Leach's Petrel. Anat. Rec. 105:110,1949.

Folk, G.E., Jr. and Peary, R. E., Jr. Experiments on the physiology of foot perspiring. CRL Special Report No. 37. May 1950.

Folk, G.E., Jr. and Peary, R. E., Jr. Penetration of water into the human foot. Report No. 131, Climatic Research Laboratory, Natick, Massachusetts, pp. 1-44,1951.

Folk, G.E., Jr. Observations on the body temperature of Leach's Petrel. Anat. Rec. 111:125,1951.

Folk, G.E., Jr. Response to physiological solutions of the isolated heart of the brown trout (Salmo trutta). Anat. Rec. 117:638,1953.

Folk, G.E., Jr. The reduction in accumulation of perspiration by impermeable barriers worn next to skin. Proc. Int. Physiol. Congress XIX:354. 1953.

Folk, G.E., Jr. A test for the penetration of Ca 45 through the integument of hairless mice. Anat. Rec. 117:599,1953.

Korean War Statistics: Dept of Veterans Affairs, http://www.va.gov/oaa/pocketcard/korea.asp. (accessed 3-16-11)

Ricklefs, RE and Rahn, H. The Incubation Temperature of Leach's Storm-Petrel. Auk 96:625–27. 1979.

## *Ch. 8 (University of Iowa)*

Berberich, J.J., Dodge, C.H., and Folk, G.E., Jr. Overlapping of ranges of eastern and western hognose snakes in southeastern Iowa. Proc. of Iowa Acad. Sci. 78:25-26,1971.

Dodge, C. H. The musk turtle in Iowa. Herpetologica 12: 176; 1956.

Dodge, C.H. and Folk, G.E., Jr. Notes on comparative tolerance of some Iowa turtles to oxygen deficiency (hypoxia). Iowa Acad. Sci. 70:438-441,1963.

Dodge, C.H. and Folk, G.E., Jr. A case of rattlesnake poisoning in Iowa. Proc. Iowa Acad. Sci. 67:622-624. 1960.

Essler, W. O. and Folk, G.E., Jr. The determination of 24-hour physiological rhythms of unrestrained animals by radio telemetry. Nature 190:90-91, 1961.

Essler, W. O., and Folk, G.E., Jr. A method of determining true resting heart rates of unrestrained mammals by radio telemetry. Animal Behavior 10:168-170,1962.

Folk, G.E., Jr., Dodge, C.H. and Folk, Mary A. Resting body temperature of raccoons and domestic rabbits. Proc. Animal Care Panel. 7:253-258, 1957.

Folk, G.E., Jr. 1966. *Introduction to Environmental Physiology*. Philadelphia: Lea & Febiger. 311 pp. (Reprinted in 1969).

Richardson RL, Fisher AK, Folk GE. Dental tissues of wild and laboratory-raised hibernating and non-hibernating 13-lined ground squirrels. J Dent Res September 40: 1029-1035. 1961.

## *Ch. 9 (Alaska)*

http://www.aeic.alaska.edu/quakes/Alaska_1964_earthquake.html

http://fairbanks-alaska.com/arctic-circle.htm

https://www.cia.gov/library/publications/the-world-factbook/index.html

Brewer M, Schindler JF. 1997. Native Naturalists. In: Science in the Community: 50th Anniversary of the NARL, p. 9-11. Barrow, Alaska.

Folk, G.E., Jr. Daily physiological rhythms of carnivores exposed to extreme changes in Arctic daylight. Fed. Proc. 23:1221-1228,1964.

Folk, G.E., Jr. Simmonds, R.C. and Folk, M.A. Physiology of deep hypothermia of the black bear. 16th Alaskan Sci. Conf., AAAS, 16:30-31, 1965.

Hedge, R.S., Folk, G.E., Jr. and Brewer, M.C. Studies on winter lethargy of black and grizzly bears. 16th Alaskan Sci. Conf., AAAS, 16:30-31,1965.

Folk, G.E., Jr. Physiological observations on subartic bears under winter den conditions, pp. 75-85. In Mammalian Hibernation III. K. Fisher et al., eds. Oliver & Boyd: Edinburgh, Scotland, 535 pp., 1967.

Folk, G.E., Jr. Physiological research in northern Alaska. Arctic (Journal) (Dedication Volume) 22:315-326,1969.

Folk, G.E., Jr. Description of the Institute of Arctic Biology. Bull. of Hibernation Infor. Exchange 90 (Feb.):1-2, 1969.

Folk, G.E., Jr. Brewer, M.C. and Folk, M.A. Cardiac physiology of polar bears in winter den conditions. Proc. Internat. Biometeorol. Congr. 5:4, 1969.

Folk, G.E., Jr. 1974. *Textbook of Environmental Physiology, 2nd Edition.* Philadelphia: Lea & Febiger. 468 pp. (Reprinted 476 pp., 1985)

Folk, G.E., Jr., Larson, A and Folk, MA. 1976. Physiology of hibernating bears, p. 373-381. In: Bears: Their Biology and Management, Vol. 3, Third International Conference on Bear Research and Management, Binghamton, New York, USA, and Moscow, U.S.S.R., June 1974. M Pelton et.el. (eds), IUCN Publications New Series no. 40, pp. 373-380.

Folk GE, Thrift DL, Persons CR. 2002. Alpha female wolf selects beta male in tundra pack. J Vet Adv 1(1):23-27.

Gray D. 1970. The Killing of a Bull Muskox by a Single Wolf. ARCTIC, North America. Jan. 23:197-199. Avail. at: <http://arctic.synergiesprairies.ca/arctic/index.php/arctic/article/view/3172/3148>. Date accessed: 04 Aug. 2011.

Line L. 1997. "Super bird - the snowy owl - Cover Story". National Wildlife Magazine, Feb-Mar 1997, vol 35, no 2 . Avail. at <http://findarticles.com/p/articles/mi_m1169/is_n2_v35/ai_19046732/>. Date Accessed 8/4/2011.

Species at Risk Act, 2003. Environment Canada. http://www.sararegistry.gc.ca. Date accessed 8/3/11.

## *Epilogue (Continuing Work and Life)*

Buresh CT, Folk GE, Dickson EW, and Thrift DL. 2010. Comparing Hypothermia in the Human and the Black Bear (Ursus americanus). Bio Rhythm Res 41(4):247–257 (First pub Dec 2008).

Dickson EW, Ludwig PS, Ackermann LW, Buresh CT, Denning GM. 2006. Met-enkephalin-Arg-Phe (MEAP): a cardioprotective hormonal opioid. Acad Emerg Med Aug;13(8):813–9.

Folk, G.E., Jr., Riedesel, M.L., and Thrift, D.L. 1998. *Principles of Integrative Environmental Physiology*, 502 pp. Washington DC: International Scholars Publications. (Reprinted, Maryland: University Press of America, 2000.)

Folk, GE, Dickson, EW, Hunt, JM, Nilles, EJ, and Thrift, DL. 2008. QT intervals compared in small and large hibernators and humans. Bio Rhythm Res. 39:427–438.

Folk GE w/ Thrift DL. 2010. The Harvard Fatigue Laboratory: contributions to World War II. Adv Physiol Edu 34:119-127, 2010.

Snetsinger P. 2003. "Birding on Borrowed Time," The American Birding Association.

Made in the USA
Monee, IL
04 November 2026

67387224R00138